PUBLIC SCIENCE, PRIVATE INTERESTS

JANET ATKINSON-GROSJEAN

Public Science, Private Interests

Culture and Commerce in Canada's Networks of Centres of Excellence

UNIVERSITY OF TORONTO PRESS
Toronto Buffalo London

Toronto Buffalo London
Printed in Canada

ISBN 13: 978-0-8020-8005-9
ISBN 10: 0-8020-8005-7

Printed on acid-free paper

Library and Archives Canada Cataloguing in Publication

Atkinson-Grosjean, Janet
Public science, private interests : culture and commerce in Canada's Networks of Centres of Excellence / Janet Atkinson-Grosjean.

Includes bibliographical references and index.
ISBN 0-8020-8005-7

1. Research – Canada. 2. Research grants – Government policy – Canada. 3. Networks of Centres of Excellence (Canada). 4. Science and state – Canada – History – 20th century. 5. Federal aid to research – Canada – History – 20th century. I. Title.

Q180.C3A84 2005 507.2'071 C2005-904228-1

This book has been published with the help of a grant from the Canadian Federation for the Humanities and Social Sciences, through the Aid to Scholarly Publications Programme, using funds provided by the Social Sciences and Humanities Research Council of Canada.

University of Toronto Press acknowledges the financial assistance to its publishing program of the Canada Council for the Arts and the Ontario Arts Council.

University of Toronto Press acknowledges the financial support for its publishing activities of the Government of Canada through the Book Publishing Industry Development Program (BPIDP).

For Stephen Straker (1942–2004) – mentor, friend, teacher, colleague – who shaped this manuscript, my thinking, and my academic life in innumerable ways. Straker was a one-off. How we miss him!

Truth and understanding are not such wares as to be monopolized and traded in by tickets and statutes and standards. We must not think to make a staple commodity of all the knowledge in the land, to mark and license it like our broadcloth and our woolpacks.

John Milton, *Areopagitica* (1644)

Contents

Figures and Tables

Figures

Tables

Preface

When the Canadian government created the Networks of Centres of Excellence (NCE) experiment in 1988, it introduced the most dramatic change in Canadian science policy since the National Research Council was established in 1916. The federal government envisaged a web of national research networks – 'research institutes without walls,' anchored in academic settings – that in partnership with the private sector would target and develop practical and commercial applications. These twin facets – practical goals and distributed networks – were unprecedented and highly controversial at the time. Now, they are central themes of contemporary science policy, and the NCE program is a central element of the government's 'innovation agenda.' Scientific excellence, commercial relevance, and public-private collaborations are recurrent themes in all new programs. Funding is targeted to areas of strategic importance to Canada's prosperity and international competitiveness. The thirty NCEs funded to date (see appendix A) 'broke trail' and initiated a fundamental shift in the organization of science in Canada. An internal NCE report captures the spirit of recent changes:

> Where previously it was difficult to find network-type programs, now we have the Canadian Institutes of Health Research (CIHR). Where once it was considered potentially dangerous to link pure research too closely to industry, now the National Research Council encourages (or at least supports) its staff to create start-up companies to exploit NRC research findings. Where 'sole-author' science was once considered the ultimate test of scientific strength, now there are programs that encourage not just national and international collaboration, but full intercontinental collaboration.[1]

Beyond its pioneer status, the NCE experiment is interesting for a number of reasons. Not least of these is the explicit intervention of the policy establishment in the culture of academic science. The radical goal (later modified) was to turn university researchers *away* from basic science and *towards* commercial application. Further, the tradition of serendipitous discovery was far too anarchic for the policy establishment; research should not only be 'managed' – a novel concept – but managed on private-sector rather than academic principles.

The design of the program is also worthy of exploration. Tensions seem to be purposive. NCEs are institutionally ambiguous in design, for example, occupying indeterminate public/private spaces, inside/outside the academy. As such, they 'float above' universities – which fund a significant portion of their costs – with little accountability to their host institutions. Other 'hardwired' stresses include commitments to *both* fundamental inquiry *and* the exploitation of intellectual property; private-sector investment *and* public funding; academic ideals *and* commercial values. This abundance of novelties would seem attractive to anyone interested in the shifting terrain of science, economics, and policy. Yet, surprisingly, the program has largely escaped scholarly notice.[2] Hence this book and the study that supports it.

A Note on Qualitative Evidence and Reporting

Case-study research methods of the type used in this book draw on multiple sources of evidence: the documentary and historical record, 'hard' quantitative data such as statistical and financial reports, and 'soft' qualitative evidence derived from observations and interviews. To assist readers who may not be familiar with the analysis and reporting conventions surrounding the latter category, the following brief overview is offered.[3]

A qualitative study must accommodate the subjective perceptions of both researchers (observer bias) and participants (response bias) within the research frame. The world views, values, and interpretations of researchers act to filter results and shape what is reported. At the same time, open-ended interview questions may generate socially acceptable responses, or a version of what respondents think researchers want to hear (deference and expectancy effects.) Such biases are difficult to eliminate completely and can lead to misinterpretations. However, certain techniques are available that help control for validity.

The first line of defence is a reflexive researcher, willing to question herself or himself and the accuracy of the data on a constant basis. The second line of defence is a set of corroborative processes, collectively known as *triangulation,* through which researchers constantly check and double-check findings. To triangulate is to compare and contrast results derived from different *sources,* such as persons, times, and places; *methods,* such as observation and interview; *researchers,* where a team approach is used; and *data types,* such as qualitative texts, tape recordings, and quantitative results. Triangulation highlights interesting contradictions, which must be pursued and clarified. Most of all, it supports the researcher's identification of convergences (themes and patterns) across multiple sources and types of data, adding confidence to her interpretations.[4]

The themes and patterns that emerge are analysed, interpreted, and reported. In the reporting process, researchers face the particular challenge of selecting, from the mass of materials they have collected and analysed, one or two samples that will effectively *illustrate* a particular theme and interpretation. One or two quotations, for example, must *stand for* all the supporting interview evidence on that topic. One or two voices *represent* the topic as a whole. This is an important point to remember in reading the qualitative sections of this book: selected quotations represent more than one person's opinion.

Throughout the book, therefore, except where explicit permission has been sought to name and attribute, as it was with the 'core-set' of scientists studied, quotations from sources are not coded, attributed, or otherwise identified. There is always a chance that codes can be penetrated by 'insiders,' risking potential embarrassment for sources, and yet those codes would be meaningless to general readers because they have no access to the underlying records. Instead, a general descriptor is used, such as 'founder member,' 'program officer,' 'network scientist,' 'network manager,' 'merchant,' or 'settler.' This approach is congruent with the confidentiality provisions of informed consent and qualitative research ethics. The author's original manuscript retains coding, as do the transcript databases. Access to both may be obtained, on request, by academic researchers in this field.

Without the generosity of my sources, there would be no study. I want to thank the scientists, policy advisers, administrators, bureaucrats, and many others who contributed their time and reflections to help me understand their complex worlds.

Abbreviations

ACST	Prime Minister's Advisory Council on Science and Technology
ANT	actor-network theory (aka translation sociology)
'big pharma'	the multinational pharmaceutical industry
CBDN	Canadian Bacterial Diseases Network
CGDN	Canadian Genetic Diseases Network
CHEO	Children's Hospital of Eastern Ontario
CIAR	Canadian Institute for Advanced Research
CIHR	Canadian Institutes of Health Research
CMMT	Centre for Molecular Medicine and Therapeutics
CSA	Canadian Standards Association
HSC/UT	Hospital for Sick Chldren/University of Toronto (*and see* Sick Kids)
HSJ/UM	Hôpital St Justine/Université de Montréal
HUGO	International Human Genome Organization
ILO	University-Industry Liaison Office (also known as Technology Transfer Office; Commercialization Office)
IP	intellectual property
IPR	intellectual property rights
IRAP	Industrial Research Assistance Program (see NRC)
ISTC	Industry, Science and Technology Canada (now Industry Canada)
MOSST	Ministry of State for Science and Technology
MRC	Medical Research Council
NABST	National Advisory Board on Science and Technology
NCE	Networks of Centres of Excellence
NRC	National Research Council

NSERC	Natural Sciences and Engineering Research Council
NSF	National Science Foundation (U.S.)
OCI	Ontario Cancer Institute
OECD	Organization for Economic Co-operation and Development
ORF	Ontario Research Foundation
OST	Observatoire des sciences et des technologies
PENCE	Protein Engineering Network of Centres of Excellence
PI	principal investigator
PRECARN	Pre-Competitive Advanced Research Network
PRI	Policy Research Institute
PUS	Public Understanding of Science
R&D	research and development
Sick Kids	Hospital for Sick Children, University of Toronto (HSC/UT)
SME	small and mid-sized enterprise
SPRU	Science Policy Research Unit (U.K.)
SSHRC	Social Sciences and Humanities Research Council
UBC	University of British Columbia
UC	University of Calgary
UQAM	Université du Québec à Montréal
U of T	University of Toronto

PUBLIC SCIENCE, PRIVATE INTERESTS

Introduction

For the past two decades, the Group of Eight (G8)* nations have structured their public research goals and research funding envelopes around the neoliberal orthodoxy of marketization, privatization, and public-sector reform. In such a climate, university research is expected to be commercially 'relevant' and open to the market. These expectations are institutionalized in 'strategic science' policy.

The changing status of academic research under the combined weight of state and market forces has been the topic of considerable debate in the literature.[1] Many scholars perceive a radical break with past values as systems of 'knowledge production' respond and adapt. Distinctions between public and private domains, and between basic and applied knowledge, are widely perceived as collapsing. Concepts such as 'academic capitalism' and 'the entrepreneurial university' attempt to capture the phenomenon. The academic life sciences are in the forefront of this revolution. The attention paid by the media to 'big hit' discoveries, gene patents, and life science venture funds helps cement the perception that the academic entrepreneur – complete with portfolio of patents and start-up companies – is the new norm.

But is this the case? Given the fundamental nature of the shifts in question, it seems important to address the question empirically. How are macro-level changes in policy and ideology playing out at the micro-level of the university or hospital laboratory? In what ways are academic bioscientists responding to the promise of the market and the

---

*The G7 before Russia's entry in 1997. Members: Canada, France, Germany, Italy, Japan, Russia, U.K., U.S.

economic imperatives of the 'knowledge-based economy'? How has the research culture changed after two decades of strategic science policy?

In the Canadian case study reported in this book, only a small minority ('merchant scientists') embraced the commercial turn. Of these, most were engaged in what I have come to call 'adventures in the nature of trade.' This archaic and useful term describes the in-between kinds of enterprise that play around the edges of commerce without fully embracing it.[2] The majority ('the settler class') largely rejected commercial ambitions and the ethical conflicts of interest and commitment that often accompany these ambitions. Subsequent inquiries have tended to confirm the relative invariance of the merchant-settler ratio. I argue, therefore, that despite the high visibility of merchant science and the powerful ideological and policy drivers that promote it, the values of the settler class (i.e., open inquiry and the free flow of ideas) still dominate academic science. The book calls into question external attempts to transform the research culture and offers the merchant/settler distinction as a useful heuristic for thinking about these questions. It is not a dichotomy, however. The book points to the importance of 'translational research' – the work that moves discoveries from the context of the laboratory to the context of use – as the linking mechanism between the two cultures.

Policy Context

Elsewhere, I have used the term 'international ideas' to describe the way in which novel combinations of principled and causal beliefs gain widespread currency and become entrenched as policy. I argue that new ideas tend to circulate first among international 'knowledge elites' of policy professionals in international organizations such as the Organization for Economic Co-operation and Development (OECD), the G7/G8, and the World Bank. These expert communities then disseminate the concepts to member states, creating convergence around particular regimes or models.[3]

Starting in the early 1980s, states began to converge around a comprehensive, ideologically driven withdrawal from Keynesian attempts to manage the economy. Governments of all political complexions embraced neoliberal ideals of fiscal restraint, tax minimization, deregulation, and marketization. States divested themselves of public utilities, nationalized industries, national airlines, and controlling interests in key industries.

At the same time, states adjusted their redistributive functions. Here, too, the logic of the market prevailed. Citizens were to become self-regulating 'enterprises' and market themselves accordingly. Translated into the public service, this reformist spirit became known as the 'enterprise' or 'entrepreneurial' model, more formally 'New Public Management' (NPM). This new culture took as axiomatic market-like principles of cost recovery, competitiveness, and entrepreneurship in the provision of public services. At the same time, accounting, auditing, and accountability measures normalized the new principles and entrenched them in the public-service ethos.[4]

From the 1980s on, then, the public funding of academic science began to be contingent on these same principles. States converged around a recognition of the economic importance of scientific knowledge and sought the best ways to harness academic science to the economy. A rhetoric of 'innovation' and 'systems of innovation' entered the policy discourse at this time. Arie Rip has proposed the term 'strategic science policy regime' as a descriptor for the mechanisms that became embedded during this period and that still prevail. On one hand, the state 'steers' towards goals of science-driven economic development through 'more or less stabilized rules of how to proceed.' On the other, an emergent new scientific establishment 'rows' towards the same goals, 'forging new alliances with policy makers and societal actors on this basis.'[5] The broad outlines of strategic science in Canada emerged as part of this general internationalizing moment.

The neoliberal agenda first came into play in Canada in 1984, with the election of a Progressive Conservative government. Science, technology, and innovation were a high priority for the new government, but Canada had never had a national science policy. Correcting this gap became a key focus of reform measures. Federal-provincial agreement was achieved in 1986, and the new policy was announced in early 1987. Creation of a 'superministry' – Industry, Science and Technology Canada (ISTC)[6] – clearly signalled the alignment of science and commerce. A new flagship strategy – the Networks of Centres of Excellence (NCE) program – helped launch the new policy. The NCE program pioneered a university-based system of national research networks and academy-industry partnerships to target and develop commercial opportunities. A unique feature was the role of professional administration – the NCE program was the first funding program to incorporate managerial requirements. An explicit goal was to change the research culture itself, moving basic scientists along the continuum to application. The program was made permanent in 1997, and it remains a key feature of Canadian science policy. The Canadian Genetic Diseases Network (CGDN), inaugurated in 1990, was one of the first NCEs funded. CGDN provided the empirical context for the two-year case study (1999–2001) on which this book relies.

Outline of the Study

The research reported in this book constitutes the first full-length, academic analysis of the NCE program. It takes a dual approach, first constructing a detailed 'top-down' narrative of the program's development and situating it in the political, economic, and policy context. Sources for this section include the policy makers and civil servants involved in launching and maintaining the program, as well as the documentary trail any such initiative leaves behind.

A second level offers a detailed, 'bottom-up' case study of the Canadian Genetic Diseases Network. CGDN brought together medical genetics researchers across the country, under the leadership of Scientific Director Michael Hayden. By the end of my fieldwork, in 2001, the network had grown to encompass some fifty scientists, eleven universities and hospitals, and eight companies. Support received from the NCE program to that point totalled $50 million. The network's research program covered four integrated themes: gene identification, pathogenesis and functional genomics, genetic therapies, and genetics and health care. 'Core facilities' in major centres offered platform technologies, such as DNA sequencing, genotyping, and bioinformatics training.

CGDN's 'child' organization – the Centre for Molecular Medicine and Therapeutics (CMMT) – opened in Vancouver in 1998. Merck Frosst, a founding industry partner, contributed $15 million towards the centre. The network's commercial profile can be seen in the major intellectual property (IP) agreement it brokered between Schering Canada Inc. and the University of Toronto; at the time, this was the largest university IP agreement in Canadian history. The agreement was based on the 1995 discovery, by a network researcher, of two genes for early-onset Alzheimer's disease. Schering's initial $9 million contribution represented seed funding for a research program in the development of drugs and technologies to treat and prevent Alzheimer's. Over the long term, the agreement has a potential value of $34.5 million, not including royalties.

In 1997, the NCE program announced that funding for all networks would be 'capped' at fourteen years. This announcement resulted in a major strategic shift for CGDN. In 1998, the network incorporated itself as CGDN Inc. It adopted a corporate organizational form and an aggressively commercial focus. The goal was to maximize revenues from licence agreements and equity holdings to replace the $4.5 million a year in federal funds that would cease in 2005, when the network reached its fourteen-year cut-off. Many of the network's scientists were ambivalent about the direction being taken. They accepted that action was needed if the network was to survive but regretted the loss of collegiality and openness that had marked earlier days.

If one imagines public and private domains and basic and applied science as intersecting dimensions (see figure I.1), CGDN had, until this point, concerned itself predominantly with 'public science' and 'basic research.' Approximately 70 per cent of NCE core funding supported fundamental, discovery-based research, and 20 per cent went to the early-stage development of technologies with commercial potential; the remaining 10 per cent supported networking and administration. But the goal of sustaining the network beyond 2005 accelerated a shift to the 'private science' half of the matrix.

A key question for the study was how this policy shift towards 'private science' affected the public's social and economic return on investment in CGDN and NCEs more generally. The tension between the public and private faces of NCEs became increasingly apparent over the course of the study. Alternating currents of confidentiality and openness ebbed and flowed around the project. Scientists spoke to me freely, while gatekeepers erected barricades. The contradictions and barriers to access illustrated the normative and ethical boundaries that are constantly negotiated in these networks.

Barriers to Access

Federal Gatekeepers

The vast majority of people who designed and implemented the NCE program in the mid-1980s had disciplinary roots in the sciences. Most held PhDs and were associated with the research councils. In interviews, their commitment to a scientific culture of openness prevailed. In contrast, 'career bureaucrats' were generally guarded; they refused

Figure I.1: Conceptual matrix for scientific research

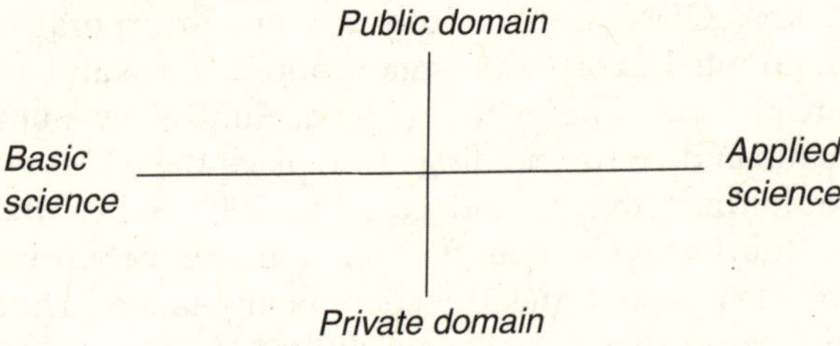

to provide key materials, citing commercial and/or cabinet and/or third-party confidentiality. Even though the research was funded by a fellowship from the Social Sciences and Humanities Research Council of Canada (SSHRC), one of the three federal granting councils, access was formally denied.

Although the wording of the formal denial sidestepped an outright claim that NCE files were exempt from disclosure, it was made clear that access would require implementation of the provisions of the Access to Information and Privacy Acts. The nature of the problem is demonstrated in the following extracts from correspondence with the program secretariat:

> Because of the sensitivity of the files in this case, we really have no option but to 'do it by the book.' This means that in order to gain access to documents in NCE files, we will have to ask you to submit formal Access to Information Act requests ... Many documents within [NCE files] would have to be reviewed on a line-by-line basis to identify information subject to exemption. And in many instances a decision about the operation of a particular exemption could only be made in consultation with other federal institutions, with networks, and with any other parties affected by the disclosure.[7]

As a result, each requested document would have to be identified by name. Not only was this quite impossible to do without the researcher having prior access to the files; it would also have made the delays and costs unmanageable. Canada's information commissioner has criticized precisely this type of strategic use of the Access to Information Act by public servants. He referred to 'the stubborn persistence of a culture of secrecy in Ottawa' and complained of too-frequent recourse to claims of third-party 'commercial sensitivity' to avoid the release of documents.[8] When public information disappears behind a privacy screen erected by public servants, questions are bound to be raised about accountability and the abuse of power.[9]

Network Gatekeepers

At the network level, within CGDN the contrast between the exaggerated discretion of professional staff and the openness of network scientists was marked. And the balance of power between scientific officers and professional staff appeared to be highly delicate, given CGDN's goal of commercial sustainability by 2005. CGDN's scientific director, Michael Hayden, belonged to both cultures. His instincts were to be open, but his position required him to attend to the concerns of staff.

As the network's scientific director, Hayden was the first person interviewed in the pilot study and the person to whom the subsequent request for access was addressed. He was an enthusiastic supporter of the idea of participant observation and committed the network to full cooperation. An internship was envisaged that would provide total immersion in network affairs. But despite Hayden's unwavering endorsement, professional staff at the CGDN's administrative centre refused access. Commercial sensitivity was the formal reason given by network management: 'Many of our interactions with industry involve the element of confidentiality, and an "outsider" may impact those discussions negatively.'[10]

Hayden's repeated interventions were required before a compromise was reached and limited access granted. There would be no internship, and access to materials was curtailed. Board and committee minutes and commercial files were not open for examination. None of the other materials could be photocopied or removed. Attendance at board, committee, and staff meetings was not permitted. Only information that staff considered 'in the public domain, i.e. financial and annual reports ... funding proposals and interim reports' would be provided.[11] By the time fieldwork was initiated at the network's head office, fifteen months had elapsed since the initial application for access.

Despite repeated requests, board and committee minutes were never made available. Eventually, a written explanation was requested. The response provides an elaborate rationale for considering aspects of this public entity as private: 'Management has received a legal opinion recommending against public disclosure of Board Minutes. The CGDN Board is a legal entity and as such, holds the right to maintain confidentiality of its in-camera meetings. Management does not hold the right to disclose those proceedings.'[12] The argument helps illustrate a key difference between NCEs and traditional university research centres. Despite public funding and public location, their governance structure is private.[13] Such tensions between the public and private faces of taxpayer-funded research recur throughout the study and will be explored in detail later.

Interviews with private-sector and researcher board members were extended to compensate for lack of access to the documentary record. Evidence of board and committee discussions and decisions came to light in other materials, such as those prepared by the network for NCE site visits.

Overview of This Book

This book inquires into the material and epistemic spaces of the NCE program in general, and of CGDN and its scientists in particular. Its detailed descriptions of the social, cultural, and material mechanisms at work in both the program and the network draw their authenticity from the voices of federal public servants, network officers, private-sector partners, university administrators, and scientists themselves. The book traces the trajectory of the NCE program and CGDN over time, showing how federal policies are translated into specific research projects, practices, and institutional arrangements and recording how scientists embrace, resist, or ignore these initiatives.

The research methods used in this project (see appendix B) borrow from the interdisciplinary traditions of science studies. Science studies is a broad church embracing many sects, including three that inform this book: micro-studies of laboratory and organizational practices, the economics of science, and science policy studies. The research incorporates a case study of the ways in which individual and institutional actors construct, extend, stabilize, and maintain complex networks and power relations: the focus of actor-network theory (ANT). The characteristic descriptive vocabulary of ANT will be deployed at particular points. The text will attempt to clarify terminology in the course of use.

The study's *primary* contribution is empirical: collecting and systematizing data on the Networks of Centres of Excellence (NCE) program and the Canadian Genetic Diseases Network (CGDN) in the context of

Canadian science policy. Chapter 1 extends the discussion of where the public interest is located in the public/private, basic/applied matrix. It focuses on the fundamental tension between 'open science' and proprietary knowledge and sets up the two conceptual models that guide the study. It then discusses some useful analytical tools, including actor-network theory and science studies more generally. Chapter 2 addresses the historical and structural factors contributing to the development of the NCE program.

The CGDN case study begins in chapter 3, which describes the network-building activities of this group of medical geneticists and the institutional identity they constructed. Chapter 4 demonstrates the way in which the network evolved a culture and sense of community, critically examines the rhetorical construction of the network's research program, and points to the authentic locations of 'network science.' These two chapters represent the public face of the network; the next two chapters move to the private side of network identity.

Chapter 5 describes the trajectory from public to private and from basic to applied in terms of the network's development of intellectual property and construction of a commercial portfolio. Chapter 6 contributes to theory by developing an analytical framework for classifying network researchers based on their alignment along the public/private and basic/applied dimensions. Chapter 7 brings together key themes to support a detailed interpretation of the costs and benefits of the NCE program and strategic science policy more generally.

The purpose of this book is to achieve a greater understanding of changes in the organization and motivation of academic science as well as of the ways in which these changes affect the public's manifold interests in the science it funds. Beyond the book's academic contributions, it is hoped that the close examination of how *science* is planned and produced and of the ways in which the *public interest* is served in this particular case will contribute to the goal of transparent, evidence-based policy in Canada.

1 Two Divides

The conceptual framework of this study relies on the relationship between two sociological and epistemological distinctions: the public/private and basic/applied divides, respectively. The space where these dimensions intersect is particularly relevant to this study, and various theorists have attempted to describe it. Donald Stokes, for example, calls the space Pasteur's Quadrant; others speak of 'strategic research,' 'emergent science,' or 'Jeffersonian science.'[1] Two models present opposing interpretations of the relation between the divides: the *open science* model and the *network* (or *overflow*) model. The tension between these contrasting approaches to public and proprietary knowledge runs throughout this book. The last part of this chapter introduces the analytical tools that are useful in understanding the conduct and culture of science in NCEs.

Mapping the Divides

Public and Private

The public/private demarcation is one of the core distinctions in sociology; it has been referred to as one of the 'grand dichotomies' of Western thought.[2] But on closer examination it begins to collapse, becoming not one but a number of related oppositions that nest one within the other like Russian dolls.[3] Is the stock market, for example, public or private? From one perspective it is a mass of individuals pursuing private interests, but from another it is a public social and cultural aggregation. What do we mean by 'the private sector'? Usually, the term refers to private businesses, large and small. Yet many of the largest corporations are 'public companies,' owned by millions of shareholders, some individual, some huge and institutional. But huge institutional investors are themselves often 'public,' in that they represent the pensions and investments of millions of people.

What is 'the public sector'? Many publicly owned institutions and agencies are 'private' in the sense that they are exempt from direct, or even delegated, public control; for example, crown corporations, universities, and even departments and bureaucracies of the state. What does it mean to speak of 'public' and 'private' life? For individuals in 'public life' we designate whole areas exempt from public scrutiny (private matters of conscience, conviction, family, and morality). But when these aspects of private life impinge on or attract the public interest, they enter the public domain and become 'public knowledge.'

Does my body belong to me? If so, I should be able to control what happens to my genetic material. But legal cases have been fought and won by researchers who have taken cell lines from unsuspecting patients and patented them for profit, rendering bodies 'public' by acts of privatization. On discovering that their genetic material had been appropriated, patients fought not for the right to privacy but for the right to profit for and from themselves.[4]

What about ownership of the human genome? In the vast undertaking to map it, public researchers raced against a private company (Celera Genomics), which sought to patent and profit from 'the stuff of life.' Because results of the public effort were held in common in the public domain, Celera was able to use them to advance its own project. The controversy raised awareness of the role of patent law in privatizing public research. Patents make knowledge private by circumscribing ideas with property rights; so, if a public university takes out a patent on a publicly funded discovery, is it 'privatizing' that knowledge? Or is it securing the ownership of that discovery for the public domain? As well, the patent process helps make otherwise private knowledge public. The alternative is to keep an innovation private by maintaining 'trade secrets,' as many companies do.

These questions without answers help to illustrate that the public/private demarcation is a negotiated, discursive space rather than a fact of the world. But two core ideas – accessibility and commonality – help connect the many different meanings. These are, as Paul Starr says, 'that public is to private as open is to closed, and that public is to private as the whole is to the part.'[5] In the first sense, the openness and transparency of public space, public life, and public disclosure are in contrast to the opaqueness and concealment of private space, private life, and personal communications. In the second sense, 'public' is synonymous with 'common,' as in public opinion, public health, or the public interest; this sense has merged with the sense of 'official' or 'state.'

Thus, for Starr, 'public' can have three contrasting meanings, from which 'privatization' represents corresponding withdrawals. In the first sense, 'public' means open and visible, as in public life and social relations, whereas 'private' means a withdrawal from sociability and the decline of public culture. In the second sense, we invoke the 'general public' or the public-at-large to speak of public action and civic concerns in contrast to private concerns and the pursuit of self-interest. The third sense of 'public' is the domain of common (state or community) ownership, as opposed to appropriation by an individual or group. These themes will reappear throughout this study.

The origins of the public/private distinction can be found in Greek and Roman thought. It represented the separation of the private household and its economy (*oikos*) from the sphere of collective public institutions – the *polis* or *res publica*. Collectively, heads of households constituted the 'body politic' or public realm. As Hannah Arendt explains, a physical space, a boundary or no-man's-land, separated private households. The boundary demarcated one property from another and marked off the household from the city. Arendt identifies the spatial significance of this boundary with that of the law. In the same way that the law harboured and protected the public domain that was political life, fences sheltered and protected the private property of households. Between the political (public) and intimate (private) domains, Arendt interposes a third space: that of the social.[6]

By feudal times, public/private distinctions in property and affairs had developed a certain taxonomic and ideological slipperiness. As historian Toby Huff describes, the emerging concept of the corporation under Roman and canon law is a case in point. A corporation interpolates between the individual and the collective, the political and the economic; in a sense, it is both public and private and neither public nor private. After the church assumed a corporate mantle to sever itself from state control, the principle of incorporation spread into secular law, where it established the rudiments of a public sphere free of ecclesiastical control. Thus, according to Huff, 'We find in the 12th and 13th centuries the widespread emergence of a vast array of legally autonomous [corporate] entities that were bestowed with a composite bundle of legal rights and which presumed the legal authority of jurisdiction, that is, legitimate legal authority over a limited territory or domain.'[7] These newly incorporated (literally, embodied) entities included cities and towns, merchant guilds, charitable organizations, professional associations, and universities.

In Huff's view, corporations subsequently contributed to the rise of the public sphere by facilitating the extension of trade in the high Middle Ages. The original trading companies were extensions of the private economy of the family: assets and investments entrusted to the company were commingled with family assets. The developing legal theory of the corporation made it possible to disentangle familial and business affairs, implementing a distinction that converted what was previously private (*oikos*) into a public entity (the market). Huff argues that corporate law made it possible to differentiate between individuals and the corporate body. The corporate collectivity was construed as a single, legal person. A distinction now existed between ownership and jurisdiction, especially concerning assets, liabilities, and debts. By providing for allegiance to the corporation rather than to individuals, the distinction ensured the continuity of the enterprise. The historical development of these concepts, according to Huff, provided for the emergence of distinctive public and private spheres of action and interest. This separation laid the foundation for the emergence of modern science as a 'public' institution within 'public' universities by establishing a 'neutral space' of thought and action. As Huff explains,

> The medieval intellectual élite of Europe established an impersonal intellectual agenda whose ultimate purpose was to describe and explain the world in its entirety in terms of causal processes and mechanisms. This disinterested agenda was no longer a private, personal, or idiosyncratic preoccupation, but a publicly shared set of texts, questions, commentaries, and in some cases, centuries old expositions of unsolved physical and metaphysical questions that set the highest standards of intellectual enquiry ... A disinterested agenda of naturalistic enquiry had been institutionalized ... It thereby laid the foundation for the breakthrough to modern science.[8]

Scientific activity after the Renaissance occurred in relatively small, interdependent communities of practice where scientific advance rested on the veracity of individuals. As sociologist Steven Shapin describes, it depended on a culture of *honour,* epitomized by the position within the social order of the seventeenth-century English gentleman-scientist. According to Shapin, the production of scientific knowledge was, and remains, a moral enterprise built on mutual trust. Personal trust is the 'great civility' and the currency of an 'economy of credibility' in the conduct of science:

> Within such small interdependent groups as the 'core-sets' of specialized scientific practices, the economy of credibility is likely to flow along channels of familiarity. The practitioners involved are likely to know each other very well and to need each other's findings in order to produce their own. Here ... the pragmatic as well as the moral consequences of distrust and skepticism are likely to be high.[9]

Thus, trust in the public institution of science rests on trust in the private morality of its individual practitioners. The 'public' nature of scientific knowledge rests on the collective construction of a collective good, under conditions requiring reliance on the work of others. Within this 'moral economy of truth,' public and private, scientific and social, become inseparable.

Similarly, Jurgen Habermas conceives of the public sphere as a social space, first emerging in seventeenth-century English coffee houses and salons, where 'private' individuals came together to engage in rational-critical debate and thereby further the 'public' interest. Habermas distinguishes this 'authentic' public sphere from the realm of state or 'public' interests. The authentic public sphere is a dimension of private life: 'a public of private people' who come together to further the 'common good.' In Habermasian terms, however, the common good and the public sphere itself are undifferentiated.[10]

Basic and Applied Science

In the science of the seventeenth century, questions of public and private first began to map onto distinctions between basic and applied science. These distinctions are part of an ancient argument that has its roots in the classical differentiation between *theoria* and *praxis* in early Greek thought. The path of *theoria* travels from Plato through Descartes and Newton; the path of *praxis* from Aristotle through Montaigne and Bacon.[11]

The rationalism of early modern science slowly came to dominate the experiential and empiricist values of Renaissance humanism. For sixteenth-century humanists, the central demand was that thought and conduct should be *reasonable*, rather than rational – that is, tolerating social, cultural, and intellectual diversity. But after the Enlightenment, says Stephen Toulmin, ideas became 'decontextualized': scientists began to conduct careful and systematic searches for the abstract universal laws through which God governed nature.[12] In contrast, a fundamental feature of Francis Bacon's critique of institutionalized scholarship in the sixteenth and early seventeenth centuries was scholarship's ignorance of the concerns of industry and commerce, the crafts and trades. Consequently, an important part of his call for reformation involved bringing the two together so that in the reformed academy 'the sounds of industry' would be heard 'at every hand.'[13]

The contrast pair of 'pure' and 'applied' science came into general use towards the end of the nineteenth century. Thomas Henry Huxley had an aversion to the distinction:

> I often wish this phrase 'applied science' had never been invented. For it suggests that there is a sort of scientific knowledge of direct practical use, which can be studied apart from another sort of scientific knowledge, which is of no practical utility, and which is termed 'pure science.' But there is no more complete fallacy than this. What people call applied science is nothing but the application of pure science to a particular class of problems.[14]

Huxley was, however, making a nice distinction, ignoring the fact that 'technology,' particularly in industry, had its own distinct history and trajectory. Other scientists recognized the linkages between 'pure' and 'applied' inquiry, or disputed the proper place of each. As early as 1840, Prussian chemist Justus Leibig sought to establish a university program that would combine the search for pure knowledge with production training for students; he was strenuously opposed by faculty. A number of late-nineteenth-century German initiatives occurred to link the demands of the pharmaceutical industry with the interests of academic science, first through consulting and contracting arrangements, then through the establishment of independent institutes. In the United States, Thorstein Veblen was complaining about too-close relations between university laboratories and local industries as long ago as 1918. Well-documented debates from the interwar years show that conflicts of interest and commitment were not uncommon. There were disputes about intellectual property ownership and concerns about the proper role of the university. The perception of similar problems today seems to indicate continuities rather than substantiate claims that a radical break in moral and organizational culture is occurring in the scientific world.[15]

The terms 'pure' and 'applied' dominated the discourse until the 1930s, when 'fundamental' came into occasional use to avoid the moral connotations of 'pure.' Subject matter (e.g., theoretical or applied physics), rather than the motivation of the scientist, defined what was pure or applied. The phrase 'basic research' was first coined by Julian Huxley as part of a typology in which 'pure' and 'applied' each consisted of two categories: pure included 'background' and 'basic,' and applied might be 'ad hoc' or 'development.'[16]

Huxley and other British socialists, including John Desmond Bernal, were inspired by the apparent success of 'planned' Soviet science. In his 1939 book *The Social Function of Science,* Bernal advocated state steering of science through socio-economic controls and goals. In contrast to this image of social engagement, Michael Polanyi and others who opposed 'Bernalism' founded the Society for Freedom in Science to defend the ideal of a 'pure science,' a science unfettered by social constraints. According to Polanyi, 'You can kill or mutilate the advance of science [but] you cannot shape it'; any practical benefits are incidental and unpredictable. Bernal and Polanyi's dialogue on social direction and autonomy in science started a continuing debate about the relative allocation of resources to basic and applied inquiry.[17]

The same debate was underway in the interwar period in the United States, and Polanyi's position dominated. At the time, according to Roger L. Geiger, academic science was controlled by 'a tacit oligarchy of eminent scientists who shared a number of ideological convictions.' These convictions included the belief that society should support basic science because it benefited from science's discoveries. In addition, the scientists felt that funding should be reserved for the 'best' scientists, because their productivity was already established. Furthermore, determining who the best scientists might be was a matter for those scientists themselves to decide, and since government funding carried the taint of politics, private support was more preferable.[18]

Robert K. Merton captured the Polanyi zeitgeist in 1942, in his book *The Normative Structure of Science.* Merton defined pure science by its characteristic methods and institutional structure, and also by the distinctive cultural values and mores that bound the behaviour of scientists. In combination, these values and mores clearly demarcated 'science' from 'technology.' The Bernal position on socio-economic relevance was adopted by Harley Kilgore, a New Deal senator from West Virginia. Wanting publicly supported science to be politically and socially accountable, Kilgore suggested that the sole criterion for public funding should be 'manifest social utility' in the production of knowledge.

Vannevar Bush, an engineer and former president of the Massachusetts Institute of Technology (MIT) who headed the wartime U.S. Office of Scientific Research and Development (OSRD), took the Polanyi and Merton side of the debate. In *The Endless Frontier* (1945), Bush politicized Merton and Polanyi's vision of a free-standing science governed by a system of binding universal norms that underpinned the moral authority on which it rested. Adopting Julian Huxley's term, Bush described as 'basic research' what this autonomous university-based collective produced. He articulated a 'linear model of innovation' to link basic research to eventual socio-economic returns.

The dominant metaphor of the linear model is the pipeline, or 'pipe.' Fundamental discoveries are fed into one end of the pipe and move through various stages of development until they emerge onto the market at the far end of the pipe. The resultant growth fuels the economy and returns taxes, to maintain the cycle (see figure 1.1). The linear model was a powerful argument for market failure, in that *basic science* was viewed as a public good, requiring public funding to sustain it. It was argued that government investment in basic research must be preserved and that science must be left to regulate itself if the pipeline was to fuel the innovation process and produce wealth. The open dissemination of research results was a necessary condition.

In the United States, Canada, and elsewhere, these arguments provided the foundation of the postwar 'social contract for science,' a contract secured by a promissory note on the eventual but completely unpredictable technological and social spin-offs of basic science. Basic research was 'performed without thought of practical ends' and with the sole purpose of contributing to 'the understanding of nature and its laws.' According to Bush, if basic research is contaminated by premature considerations of use it loses its creative edge. But if left alone, it provides the raw materials for innovation and becomes, at a distance, 'the pacemaker of technological progress.' Thus, in the form of technology transfer, basic science generates social and economic returns on the state's investment – but only if scientists are allowed to pursue it, wherever it leads, without government controls. Government's role

Figure 1.1: The linear model of research: The Second World War to the mid-1970s

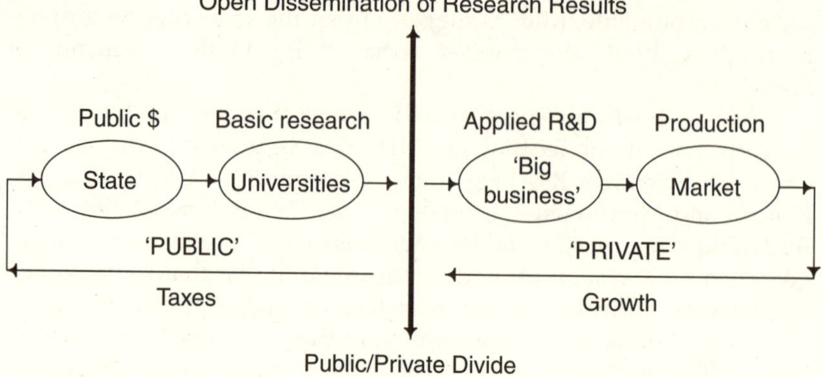

was simply to *support* university researchers with the resources they needed to produce knowledge.[19]

Many U.S. scientists viewed Bush's 'endless frontier' as 'a charter for pure science': it enshrined the basic/applied dichotomy in U.S. science policy and entrenched the 'ideology of the autonomous researcher.' Bush argued that 'the responsibility for the creation of scientific knowledge – and for most of its application – rests on that small body of men and women who understand the fundamental laws of nature and are skilled in the techniques of scientific research.' Only peers could decide the value and merit of research, so 'there was no need for governments to worry about the evaluation and measurement of science and scientists, and to track the output of research.'[20]

Politicians and policy makers initially disagreed with Bush's assessment. The U.S. National Science Foundation, for example, was not established until 1950, and then only with far more restricted authority and autonomy than Bush had anticipated. But in the late 1950s, in the aftermath of *Sputnik*, the linear explanation of the relation between basic science and application became compelling. Bush had argued that, without significant investment at the source of the knowledge pipeline, no innovations would issue from the mouth and the nation would fall behind its competitors. *Sputnik* seemed to demonstrate the truth of this claim. Fears of Soviet dominance of the space race generated immediate revisions in the U.S. federal research budget. The golden age of state-sponsored research had arrived.

The Spaces in Between

Setting up a dichotomy between basic and applied creates an artificial divide between the search for solutions to practical and technical problems and the search for fundamental understanding. As Donald Stokes argues and as the historical record suggests, basic research has never been divorced from application; distinctions between research directed to useful ends and research directed to the advancement of knowledge are deeply misguided. Stokes suggests that a large proportion of university research is – and always has been – *both* useful *and* fundamental. He maintains that the basic/applied dichotomy renders this significant segment of the research spectrum invisible. Furthermore, Stokes holds that the linear model's one-way flow obscures the number of basic research questions arising from purely technological phenomena.[21]

In furthering his claims, Stokes employs an illuminating typology. He classifies fundamental, 'understanding-based' research as Bohr's Quadrant, and applied, 'use-inspired' research as Edison's Quadrant. Research that is *both* useful *and* fundamental resides in between, in Pasteur's Quadrant. Pasteur's research commitment, according to Stokes, was twofold: not only to understand the microbiological processes he discovered but also to exert practical control over their effects in products, people, and animals: 'The mature Pasteur never did a study that was not applied while he laid out a whole fresh branch of science [microbiology].' In Stokes's view, it is this dual commitment to understanding and use that characterizes much of university research. 'Every one of the basic scientific disciplines has its modern form, in part, as the result of use-inspired basic research. We should no longer allow the post-war vision [of Bush] to conceal the importance of this fact.'[22]

In further contrast to Bush's one-dimensional linear model, Stokes sees the rise in fundamental scientific understanding and the rise in technological know-how as two loosely coupled systems. Instead of the latter being dependent on the former, each phenomenon progresses along largely independent trajectories, with no intervention from the other. But at times, Stokes argues, the mutual influences are profound and can go in either direction, with use-inspired basic research often cast in the linking role. At that point, the two trends conjoin in a seamless web. It is a commonplace that new technologies will be increasingly science based, but the under-appreciated concomitant, Stokes argues, is that science will be increasingly technology based.

Figure 1.2: Stokes's 'quadrant model' of scientific research

Research inspired by: *Considerations of Use?*		No	Yes
Quest for Fundamental Understanding?	Yes	Pure basic research (Bohr)	Use-inspired basic research (Pasteur)
	No	Research directed to particular phenomena (*Wissenschaft*)	Pure applied research (Edison)

Source: Stokes (1997)

What goes unsaid but is nevertheless clear from the discussion is the relation of 'understanding' and 'use' to 'public' and 'private.' If Bohr is public and Edison private, Pasteur occupies the shifting space *between* these two poles. The biomedical sciences today epitomize these 'spaces in between.' Physician-scientists in the case study that is featured later in this book describe much of what they do as *translational research,* a concept that fits the intermediate space between bench and bedside, laboratory and market. Policy instruments, such as the NCE program, that are geared to *both* scientific excellence and commercial relevance address research in Pasteur's Quadrant. The implications of Stokes's insight are being explored by others.[23] The model is reproduced in figure 1.2.

'Open Science' or 'Science That Overflows'?

Two opposing theoretical perspectives on academic science can lay claim to the space between basic and applied, public and private.

- The 'open science' model, grounded in the 'new' economics of science, argues that the commercial exploitation of proprietary knowledge by public universities undermines the pursuit of use-inspired basic research.
- The 'network' or 'overflow' model, grounded in science studies, argues that the genie is already out of the bottle, that institutional distinctions are largely irrelevant anyway, and that the resulting

state of affairs (intersectoral fluxes, flows, and circulations) is largely beneficial.

In 1954, Jonas Salk announced that he had developed a vaccine for polio at the University of Pittsburgh. In a television interview, he was asked why he had not taken out a patent on an invention clearly worth millions. Salk replied, 'How can you patent the sun?' His point – that no one should own or profit from discoveries about the natural world – has been overtaken by events. Patents are now used routinely to convert university research into proprietary knowledge, as part of a systematic effort to turn universities towards the market by 'capitalizing' their own research. This is where the basic/applied and public/private dimensions overlap.

The Open Science Model

University intervention in the commercialization process is highly contested on both social and economic grounds. The first category questions the social costs of commodifying public universities and their knowledge, holding that these institutions should remain outside the private system of market exchange. One argument is that although the costs of advancing basic knowledge are socialized – taxpayer supported – the benefits from its application are privatized, in the form of intellectual property rights. Some make an ethical argument that when research is publicly funded, neither researchers nor their universities have moral rights to proprietary control over resulting products.[24]

These are powerful current debates, with many facets. One of the more interesting facets is that the position of social critics is aligned, rather curiously, with the second line of contestation, which advances the economic interests of industry. This 'open science' model conceives of the new commercial role of universities and their researchers as *impediments* to industry and therefore to innovation and wealth creation. The focus on intellectual property rights creates tensions by redefining the role of universities. Rather than freely supplying ideas to the private sector, these institutions now compete with industry by 'protecting' ideas and making them increasingly costly for the productive sector to access.

Advanced as 'the new economics of science,' the open science perspective advocates a return to 'no-strings attached' public funding of basic science, a recommitment to the open publication of results, and a removal of expectations that universities should be involved in commercialization.[25] Essentially, open science seeks to 'turn back the clock' to the linear understandings of the postwar golden age discussed above, when universities produced 'public' knowledge, industry exploited it, and an arm's-length relationship kept the two sectors at a healthy distance (see figure 1.1).

Using Paul Starr's interpretation – 'public is to private as open is to closed, and ... public is to private as the whole is to the part' – the 'public' knowledge produced in universities is common property.[26] In the classic formulations of Richard R. Nelson and Kenneth Arrow, basic science is a 'public good' in the sense that the private sector will not invest in knowledge that is 'non-appropriable' and 'non-rival.' Given *market failure*, the state must correct the underinvestment in basic research. As summarized by Keith Pavitt, 'the simple economics of basic scientific research' are such that basic research generates information that is costly to produce but virtually costless to reproduce and reuse. It therefore has the properties of a public good and deserves public support. If business firms try to capture all the benefits of basic research for themselves, either through trade secrecy or property rights, knowledge remains underexplored or underexploited. Thus, state support for basic research can be justified on the grounds of economic efficiency.[27]

Yet *universities* are now patenting and licensing a 'non-trivial fraction' of what would previously have been placed in the public domain. When a university owns patents and licenses discoveries to the private sector, transaction costs for industrial development are increased because companies must now pay for techniques and materials that were previously freely available. Industry's costs also increase when university researchers spin off patented discoveries into their own companies and then license subsequent products to larger firms.[28]

Thus, transaction costs reduce accessibility. Industry prefers, therefore, to maintain university research in the public domain: 'The large pharmaceutical companies, in particular, have begun to complain vociferously that since they and the public pay for this research through taxes given to the university, it is not fair for them to pay again for access.'[29] As well, patents are said to restrict the diffusion of knowledge that promotes innovation. Traditional methods of diffusing knowledge from universities to industry – journal articles, meetings, conferences, and so on – are thought to be more efficient. Since barriers to access decrease overall wealth, arguably it is more efficient for government to subsidize the production of fundamental knowledge and give it away 'for free.'[30]

Some critics argue that although the role of the university in the knowledge economy is not yet clearly articulated, identified, or understood, inherent tensions beset *both* universities' pursuit of commercial alliances *and* their traditional 'quest for eminence.' Richard Florida and Wesley Cohen suggest that a more balanced view of the university's new role in the economy is required. Instead of positioning universities as engines of economic growth, a more nuanced perspective would reframe the university as 'an enabling infrastructure for technological and economic development.'[31] As will be seen later, CGDN has recently redefined its mission in precisely these terms.

In this vein, a 1997 empirical study of university patenting in the United Kingdom concluded that a 'regime of appropriation' in the academy, although effective in the short term, may in the medium to long term limit the overall rate of return. It argued that university patenting and intellectual property rights can unintentionally compromise the commercial potential of research, and that in securing patents the university positions itself as a potential competitor to private-sector firms. Further, university patents may hinder future development if the patent coverage has been poorly framed or prematurely filed.[32]

Disclosure restrictions on the securing of intellectual property rights may prevent research results from entering the public domain in a timely fashion. University commercialization activities may impede the cumulative advance of the research enterprise by increasing the wasteful duplication of effort and reducing the likelihood that current findings will contribute to future work. Furthermore, as a number of studies have shown, disclosure restrictions are by far the most significant economic cost associated with university patenting and licensing.[33] The need to negotiate licences on the proliferating tangle of overlapping rights significantly increases the overall costs of research and proves a major disincentive to downstream advance. The situation has been described as an 'anticommons' in that it *underutilizes* scarce resources.[34]

Restrictions in licences are pervasive. A 1997 U.S. study found that 82 per cent of companies surveyed require academic researchers to keep information confidential to allow for the filing of a patent application, and that 47 per cent have agreements with universities that allow for even longer delays. Additionally, 30 per cent reported that conflicts of interest had arisen with universities, and 34 per cent had experienced intellectual property disputes with academic researchers. The study confirmed that participation by researchers in commercialization is associated with both delays in publication and a refusal to share research results on request. Industry-supported and market-oriented biomedical researchers were more than three times as likely to delay publication as were those who had no industry support.[35]

Similarly, in a survey of technology managers and faculty at the 'top 100' R&D-performing universities in the United States, 39 per cent of managers had experienced situations where firms placed restrictions on the sharing of information between faculty. Also, 79 per cent of managers and 53 per cent of faculty reported that firms had asked for R&D results to be delayed or kept from publication. In addition to restricting the flow of knowledge, disclosure limitations also generate real and potential conflicts of interest that can damage public perception of the research enterprise.[36]

Another relevant issue for open science is the 'patent-scope' problem, the practice of taking 'broad patents' on basic biomedical platform technologies, such as recombinant DNA techniques or monoclonal antibodies. Especially problematic are rights claimed to 'whatever useful may come' from the patenting of gene sequences and fragments. The U.S. Patent and Trademark Office is beginning to demand better demonstrations of utility associated with gene sequencing, but it is still a fairly weak standard. Some critics argue that the use of broad patents to commercialize public-sector research is unacceptable, as are policies promoting such a strategy. When biomedical discoveries are converted into proprietary products, the amount of public investment required to bring them to fruition is not taken into account. Patented biotechnologies build on years of publicly funded research in 'pure' molecular biology, and the industry continues to draw on advances in public science. Richard R. Nelson claims that modern biotechnology is a canonical example of a field where science and technology, public and private, are inextricably mixed. Thus, he says, allowing those who are the first to bring discoveries to practice to privatize the whole system seems not only unfair but unjustifiable. They have simply placed the last brick on the wall.[37]

In more general terms, broad 'pioneer' patents seem to discourage further development because of the likelihood of patent infringement and the legal costs of defending such infringement. The effect is to erect a barrier around a wide area of the intellectual landscape. Nelson argues strongly that patent scope should be kept as tight as possible. To claims that broad patents are necessary to encourage inventors to innovate, he points to technologies that have been developed without such protection, such as semiconductors, transistors, and integrated circuits. He states unequivocally,

> We believe that the granting and enforcing of broad pioneer patents is a dangerous social policy. It can, and has, hurt in a number of ways ... And there are many cases in which technical advance has been very rapid under a regime where intellectual property rights were weak or not stringently enforced. We think the latter regime is the better social bet.[38]

Rather than stimulating innovation and diffusion, therefore, a tangle of fragmented and overlapping patent claims impedes the advance of knowledge. Researchers must obtain licences and pay royalties to all who hold interests in the upstream basic technologies. As a result, and paradoxically, an increase in intellectual property rights can lead to a decrease in useful products.

In 1998, in the 'Ehlers Report,' the House Committee on Science of the U.S. Congress acknowledged the chilling effect of university patenting, stating that 'a review of intellectual property issues may be necessary to ensure that an acceptable balance is struck between stimulating the development of scientific research into marketable technologies and maintaining effective dissemination of research results.'[39] In certain key respects, the Ehlers Report moderates the headlong rush to commercialization that has characterized the policy of the past two decades. Science is driven by curiosity, says the report. Its purpose is learning and discovery, and discovery requires the intellectual independence to undertake the kind of long-term research that is not immediately commercializable. Understanding-driven and 'targeted basic' research must be supported if the economy is to remain healthy. In presenting the report, the science committee chairman, James Sensenbrenner (R-WI) stated, 'The clear message of this report is that, while not exactly broke, America's science policy is nonetheless in need of some pretty significant maintenance.'[40]

Economists emphasize the continuing economic importance of sustaining basic research rather than directing it into specific and narrow commercial applications. The majority of R&D funding (80 per cent) is spent on already-existing products, that is, on improvement, not innovation. Telephones, transistors, lasers, and computers are examples of the essentially unpredictable nature of the technological outcomes of basic research investments. Basic academic research produces a multitude of new, publicly available ideas that everyone can share, thereby stimulating innovation. The enforcement of university intellectual property policies tends to choke off this important source of innovation: 'Instead of offering new and different opportunities for the Pasteurs of the university, policy makers may try to convert both the Bohrs and Pasteurs into Edisons.' Modern-day Pasteurs must continue to find a place in the university if progress is to continue. 'If badly designed policies interfere with this interaction, they can do great harm.'[41]

In summary, the conditions of knowledge production are such that the details of institutional and organizational differences between the public and private sectors 'really do matter' in the open science model. Paul David argues that the integrity of science and the scientific method depends on 'maintaining an ethos of *openness* and *cooperation* among researchers, supported by the presupposition that the *reliability* of scientific statements is a collective product requiring independent verification, and consequently conformity with some behavioural norms regarding the disclosure of their findings'[42]

As noted earlier, these institutionalist economic arguments mirror those of social critics of university commercialization, indicating a developing consensus that may be significant for future policy. But for another influential model, demarcations such as public/private and basic/applied are meaningless, and intellectual property is just one of the many 'intermediaries' in a knowledge production system constituted by flows, circulations, and network linkages.

The Network (Overflow) Model

The opposing view to the open science model, the *network (overflow)* model, argues that changes in the knowledge production system since the 1980s are radical and irreversible and constitute a productive force for good. The open science model is seen as a legacy of the long-obsolete institutions of the Cold War era. Today these institutions represent obstacles to science's ability to contribute to economic development. Especially in the biosciences and information and communication technologies (ICTs), tight coupling and multiple linkages between state policy, university research, and industry receptors is the new norm.

Public/private and basic/applied distinctions are beside the point here; what matters is the extent of the connections.[43]

This model is process based; its intellectual antecedents can be traced from Heraclitus to Alfred North Whitehead. What it attempts to describe may be closer to the historical reality than the open science model, the 'purity' of which can be seen as an artifact of postwar affluence. As suggested earlier, there was a long tradition of cross-sectoral linkages in the interwar years and before. However, the shift in degree of cross-sectoral interactions today is a marked departure from earlier times.

EMERGING AND CONSOLIDATED NETWORKS

Actor-network theory (ANT) underpins understandings of the network/overflow model. ANT attempts to account for the ever-changing networks of social and technical relations that permeate scientific facts, artifacts, and organizational arrangements. The analogy of a 'seamless web' is often used. Especially in emerging networks, the emphasis is on connection, interdependence, mutuality, and flux. Stability is tenuous, and networks can quickly dissociate without constant attention. Facts, practices, and artifacts become more settled as networks consolidate. Power and agency accrue in consolidated networks as relational effects. They enable 'action at a distance' and steering by 'remote control.' The achievement of action at a distance is exemplified by the concept of *centres of calculation*, in which the ability to control actors at the periphery of the network translates into power at the centre. The use of technologies such as printing and bookkeeping in the past, and of ICTs today, provides the necessary elements of surveillance and control. In later analysis, we will return to these concepts.

Among ANT theorists, Michel Callon's work has contributed most to debates on the relationship between science and economics. His focus on this interface makes it fruitful to engage his resistance to the 'open science' model. Callon agrees that the state should invest in basic research, and he is concerned about the confrontation between the logic of disclosure and free circulation of ideas and the logic of proprietary knowledge and secrecy. However, he rejects the economic foundations of the open science model. 'The thesis of underinvestment in research [by the market] is becoming more and more difficult to support,' he says; 'public laboratories are one after another falling into private hands, either directly through takeovers and cooperative arrangements or indirectly through incentives and research programs.'[44] Rather than defining the private domain in terms of withdrawals from the public domain, as Starr does, Callon inverts the question by pointing out that a lot of effort is required to make scientific knowledge public, whereas almost no work is required to keep it private.[45] To Callon, science has always been 'potentially privatizable'; to maintain it in the public domain requires intensive investments of energy by scientists, the state, and institutions such as universities.

Science is constructed and circulated in heterogeneous networks – *hybrid collectives* – some local and emerging, some extended and consolidated. The more networks there are, the more scientific innovation flourishes. 'Science is a public good when it can make a new set of entities proliferate and reconfigure the existing states of the world. Private science is the science that firms up these worlds, makes them habitable. This is why public and private science are complementary; despite being distinct, each draws on the other.'[46]

Local, emerging networks are private in the sense of 'intimate,' in that the space of circulation is limited. When network science overflows the local frame, the space of circulation opens up. At the same time, however, the magnitude of the investment required is enormous and tends to generate long and complex chains of associations. As the network settles into place, so the links and relations become standardized and 'heavy with norms.' This tends to produce what Callon calls 'irreversibility' and what institutional economists call path dependence and technological lock-in. In other words, the network becomes self-perpetuating and the space for the circulation of new ideas shrinks. It is at this point that intervention is needed and the hard work of keeping science public must take place. Strong, stabilized networks should receive no additional public support, says Callon. Instead, support should go to encouraging the emergence and proliferation of new networks. It is the *variety* of academic research that thwarts the tendency to lock-in. Established networks should be constrained by requirements to disclose the knowledge they produce and by limits on the duration of patent protection.

In more recent work, Callon admits that accounting for the 'dual movement' of *exploration* and *exploitation* is a difficult question, in that investments in established and profitable types of knowledge have to be encouraged at the same time as investments in new, currently unprofitable avenues of inquiry. In other words, without the incentives of 'open science,' how do we ensure a continuing supply of new discoveries? To address this question, Callon refers to fields such as biotechnology and ICT, which in his estimation successfully balance exploration and exploitation. These fields 'constitute veritable social laboratories in which new arrangements, devices, and rules of the game are tried and argued.'[47] The main problem with the open science model, according to Callon, is that it confines science within existing institutional boundaries and has little to say about the work that converts scientific knowledge into marketable applications. In other words, it addresses only codified knowledge and assumes that the same conditions also apply to tacit knowledge. Further, it assumes that we can draw clean lines between these two forms of knowledge.

In contrast, 'emerging sciences' such as biosciences and ICTs are at once autonomous *and* strongly connected to the market economy. 'Emerging sciences' seem to occupy a midpoint on the continuum between tacit/embodied and codified/consolidated. In other words, they belong in what Stokes calls Pasteur's Quadrant. Subsequent alignments stabilize emergent networks and move them towards consolidation. The reverse is also the case. Consolidated networks can unravel and cede their place to emergent networks.

MODE 1, MODE 2, AND TRIPLE HELIX

Although Callon's model is more extensively theorized, it bears a family resemblance to two formulations that began circulating in the science policy/science studies literatures in the early 1990s, when government cutbacks in research funding and enhanced expectations of commercial exploitation began to fundamentally rewrite the conduct of academic science.

In the early 1990s, Mode 2 and Triple Helix models emerged to describe the changing conditions of knowledge production. The first argues that traditional (Mode 1) discipline-based ways of producing knowledge are being replaced by new (Mode 2) transdisciplinary networks that operate along the periphery of the academy and extend beyond it. Mode 1's focus is on the context of discovery. Mode 2 extends discovery into contexts of application. The new networks bring together a variety of skill sets and different types of expertise. Unlike Mode 1, Mode 2 has an organizational structure that is flat rather than hierarchical, and structures are flexible enough to shift and recombine as the problem focus changes. A major innovation lies in accountability relations. Mode 1 holds itself accountable to the community of science, by way of peer review; Mode 2 considers itself accountable to the community at large. Quality control extends beyond traditional peer review structures to include the broader set of practitioners that populates these networks. Mode 1 may be considered analogous to Bohr's Quadrant. The focus on useful knowledge and the context of application in Mode 2 clearly suggests Pasteur's Quadrant.[48]

In a complementary fashion, the Triple Helix model proposes the recursive interaction of academy, industry, and state institutions in pursuit of knowledge-based economic development and innovation. Triple Helix proponents argue that these institutional alliances signal a new 'democratic corporatist' form that creates a new 'quasi-public sphere ... in between representative government and private interests.'[49] This new arena legitimates the state's involvement in an area that might otherwise be left to the 'invisible hand' of the market.

Integral to the Triple Helix vision is an image of a new type of university – the *entrepreneurial university*. In contrast to the passive linear model, in which knowledge is handed over to industry for exploitation, the entrepreneurial university capitalizes its own knowledge, thereby changing the dialectic between the university and society. The primary vehicles of change are public-private linkages and collaborations and dedicated structures to capture, capitalize, and exploit intellectual property. Triple Helix proponents consider these collaborations to be within the productive sector of the economy, in Edison's Quadrant.

Uncritical acceptance of the perceived dichotomy between new and traditional forms of knowledge organization would be naive. Steve Fuller speaks of 'the myth of the modes.' Far from being new, says Fuller, the 'institutional dawn' of Mode 2 and Triple Helix models occurred in nineteenth-century Germany's large-scale academy-industry-state collaborations in physics and chemistry. Positing radical breaks and new eras obscures the basic continuity in knowledge production and betrays a presentist understanding of history.[50] The new models are 'fashionable ideas,' rhetorical ploys that identify features always/already present. They favour descriptions of 'revolution' rather than 'evolution' because they are normatively predisposed towards entrepreneurial activities and public-private partnerships. Nevertheless, most OECD countries incorporate a programmatic orientation towards the new formulations.[51]

Policy Regimes

To understand how the open science and network models are being put into practice, a broad framework is required that encompasses the way the two intersecting dimensions (public/private and basic/applied) play out at the policy and program level. The concept of 'policy regimes' fulfils that requirement. Arie Rip suggests that science policy regimes manage the mutually dependent 'national research system,' a landscape made up of interactions among research performers, funders, users, markets, and state 'incentive structures.' Policy regimes lock in to particular trajectories of institutionalization. In the 1950s and 1960s, the linear model of innovation and the social contract for science dominated. In the 1970s and 1980s, a flurry of activity marked 'big science.' Today, we have the 'strategic science' regime that was initiated during the high tide of neoliberalism in the late 1980s.[52]

Finer-grained analyses can be found in discussions of 'phases' or 'paradigms' of science policy. The period since the Second World War has been divided into five 'ages' of science policy.[53]

1. The Naive Decade (1945–55): There is strong public confidence in science and scientists; policy supports basic research.
2. The Age of Pragmatism (1955–70): The overriding concern is Cold War national security; military and prestige projects prevail. From about 1960, science and technology begin to be linked to wider issues of economic growth. Resources are plentiful; investment in basic science is politically acceptable.
3. The Technology 'Cure' (1970–5): Slow economic growth coupled with energy shocks directs government attention to increased investment in technological innovation as a cure for declining growth. The idea of re-engineering the science and technology system through better university-industry linkages is developed.
4. Science as Strategic Opportunity (1975–88): Science is seen as a source of economic growth and social welfare. Scientific institutions are revamped. Science becomes something to be managed through 'networks, visions, and missions.'
5. Science as Marketplace (1985–present): Economic accountability is emphasized; science 'supply' is integrated with market 'demand' through new bridging institutions and mechanisms, multidisciplinary and multistakeholder. Distinctions are blurred between public good and private gain in the benefits of and rights to research.

Although this periodization seems generally accurate, it is debatable whether the last category, 'science as marketplace,' should be considered separately from 'science as strategic opportunity.' Arguably, both belong to the neoliberal policy regime of 'strategic science' initiated in the late 1980s.

The neoliberal ideology of the 1980s advocated a comprehensive withdrawal of the state from the economy. Regardless of political complexion, governments 'all abandoned Keynesian policies and ... pursued fiscal restraint, tax minimization, deregulation, and marketization.'[54] States began to divest themselves of public utilities, nationalized industries, national airlines, and controlling interests in strategic industries.

At the same time, states adjusted their redistributive functions. Here, too, the logic of the market prevailed. Citizens were to become self-regulating 'enterprises' and were to market themselves accordingly. Translated into the public service, this reformist spirit became known as the 'enterprise' or 'entrepreneurial' model, or, more formally, New Public Management (NPM). This new culture took as axiomatic market-like principles of cost recovery, competitiveness, and entrepreneurship in the provision of public services. At the same time, accounting, auditing, and accountability measures normalized the new principles and entrenched them in the public-service ethos. From 1980 on, then, the public funding of academic science began to be contingent on these same principles.[55]

These principles continue to dominate policy mechanisms: neoliberal science has become Arie Rip's 'strategic science.' In a strategic science regime, the state attempts to steer the research agenda and institutionalize processes of agenda building. Thus, strategic science has developed 'more or less stabilized rules of how to proceed' towards the state's goals of wealth creation and sustainability. At the same time, an emerging new scientific establishment is 'promising to contribute to [those goals] and [is] forging new alliances with policy makers and societal actors on this basis.'[56]

As Rip shows, the strategic science regime combines concerns for relevance (applied research of benefit to the private sector) with demands for excellence (basic research to enrich the public knowledge base). These ideas were 'in the air' when the Networks of Centres of Excellence (NCE) program was conceptualized in the mid-1980s. Rip speaks of 'fashions' in ideas and the 'abstract sponsorship' they exercise. Ideas matter: their power lies in their ability to provoke action. They help to 'order the world' by shaping agendas and outcomes. Modest ideas, such as relevance and excellence, and big ideas, such as new public management, systems of innovation, and the knowledge-based economy, disseminate widely and become dominant. They become 'international ideas.'[57]

In science policy, 'international ideas' are principled and causal beliefs held by international 'knowledge elites' about the economic importance of scientific knowledge and the best way to harness science to the economy. New ideas about science and the economy tend to circulate first in 'epistemic communities' of policy professionals in international organizations such as the OECD and the World Bank. These expert communities then share the new ideas with member states, creating convergence around particular regimes or models. They also help supply the formal and informal structures through which policy frameworks are negotiated and ideas disseminated.[58] The broad outlines of strategic science in Canada emerged as part of this general internationalizing movement. The material effects of international ideas can be seen in reformulated funding priorities, new infrastructures for the exploitation of intellectual property, and initiatives such as the NCE program.

'Excellence' is one of the defining tropes of strategic science. It is not an innocent term. In its fixed sense, excellence simply means high quality; this is unobjectionable. But in its relative sense, excellence means superiority, or a standard better than the norm. Understood in this manner, performers of 'excellent' research stand in contrast to a much broader population of average or marginal performers. In their critical review of the career of 'excellence' in U.K. science policy, Jordi Molas-Gallart and Ammon J. Salter champion 'mediocrity,' pointing out that 'by its very nature excellence can only be achieved by a very limited number of researchers or research groups.' These authors fear what Robert Merton called a 'Matthew effect' ('to those that have shall be given more') that would direct funding exclusively to researchers and research organizations with established records of excellence. Not only would this restrict diversity and capacity in the research system, it would cut off the important contribution of 'average' science in areas such as training the next generation of researchers, opening up new fields of inquiry, and offering a wider field of social choices about which new technologies get developed.[59]

Arguably, concentrating research funding on established scientists and institutions leads to less innovation than does spreading funds across multiple sites. Maintaining variety in the system ensures possibilities for new entrants, who often work on the margins of traditional disciplines.[60] Similar concerns about exclusion and loss of diversity were expressed when Vannevar Bush was developing 'the doctrine of basic science.' At that point, there was a fear that setting up the U.S. National Science Foundation would institutionalize 'the monolithic pressures of scientific orthodoxy' and support 'only research of a recognized kind in established fields.'[61]

In this book's later analysis of CGDN discourse, excellence will emerge as a dominant trope in the guise of a performative elitism. Inclusion and exclusion are its recurring themes. Actor-network theory provides a way of understanding how scientists and others in the CGDN continuously negotiate competing demands for excellence and relevance from the NCE program while simultaneously inventing their network. It is precisely ANT's ability to account for the workings of power in network relations that makes it such an appropriate analytical tool for this book's case study of the Canadian Genetic Diseases Network. However, like micro-studies of science in general, ANT is less helpful in accounting for the *structural* relationship between CGDN and the NCE program.

Structural Issues

Relational approaches to the study of science ask questions about the ways in which knowledge is produced at the micro level. They focus on detailed, ethnographic descriptions of local practices or on a close historical study of specific episodes. The key is simply to 'follow the actors' at the actual site of their scientific work. Explanation emerges once description has been saturated or pursued 'to the bitter end.' With such a strong focus on the local, surrounding institutions tend to become epiphenomenal 'scale effects' of relational networks. The entire research system can be viewed as a contingent *outcome* of the 'powers of association' attached to networks. Causal accounts are abandoned, and social and normative 'why' questions disappear in the minutiae of mundane 'how' questions. Political-economic issues vanish into the local politics of research. Thus, exclusively relational approaches are not only unsatisfactory but also methodologically unsound.[62]

The micro focus neglects important *existing* structural and institutional features that constrain individual and collective actors. ANT has long sought to avoid full engagement in the agency/structure debate and a more satisfactory accounting of formal institutions. It tends to fall into infinite regress when attempting to account for structural features. Peter Keating and Alberto Cambrosio frame the problem in this way:

> [T]he fact that traditional sociological dichotomies (macro and micro, social and technical, nature and culture) are inappropriate tools for describing and analyzing scientific and medical practices ... has been a leitmotif of many recent contributions to the science studies field. Yet, once the ritual rhetorical ceremony of excommunicating the usual dichotomies has been performed, the question remains of [what] analytical frame ... will allow us to move to ... an appropriate account of, say, the development of biomedicine in the last half-century.[63]

This book argues that it is possible to speak meaningfully of 'structure' – and to study its effects on the way science is organized and carried out – without reifying it. One way is to think of it as a process: 'structuring' or 'structuration,' in which actors and institutions shape each other in a continuous fashion. Another way, from within ANT, is to understand that some actor-networks become so deeply established over time that they can more or less be taken for granted as macro-social structures.[64] Perhaps this is a good enough understanding on which to base the structural (policy/program) level of this investigation. But in order to undertake the micro level of the project, we must first overcome another methodological problem.

'Studying Up'

Steven Shapin pointed out that science studies is 'one of the few sociological specialties ... that aims to interpret a culture far more powerful and prestigious than itself ... [F]ew students come equipped with relevant competencies in the natural sciences.' He calls this the problem of 'studying up.'[65] Anyone proposing to enter the social world of medical geneticists without being an initiate must deal with the issue of whether or not he or she can, or should, acquire linguistic competence in the field.

Laboratory Life, the pioneering study of scientists in action, was an 'anthropology' of science, according to its authors. The study was an ethnographic investigation, grounded in participant observation, of one specific group of scientists in one specific setting. Using anthropological means, the authors hoped to penetrate the 'closed-shop' status of science and open up scientific claims by breaking down the mystique of scientific objectivity. In order to understand the tribes they study, anthropologists usually attempt to acquire linguistic and cultural competence by immersing themselves in the field. In contrast, the *Laboratory Life* study adopted a methodological principle of 'anthropological strangeness' with regard to its subject matter.[66]

Although conducting a field-based study, the authors made a point of maintaining critical distance. They decided that an understanding of science was not a necessary prerequisite for understanding scientists' work. On the contrary, 'the dangers of *going native* outweigh[ed] the possible advantages of ease of access and rapid establishment of rapport with participants.' Thus stories of 'laboratory life' were accounts based on 'the experiences of an observer with some anthropological training, but largely ignorant of science.' In the land of science, Latour and Woolgar chose to be 'the stranger.'[67]

But to what extent can strangers, ignorant of the 'native language,' expect to penetrate the meaning of activities they observe and document? Certainly, strangers may be able to observe without bias; but on the other hand, they may utterly misinterpret what they observe. Alleged misinterpretations by science studies researchers have provided ammunition in the 'science wars.' Physicist Alan Sokal argues that these researchers' case studies are often contaminated by 'extremes of subjectivism, relativism, and social constructivism.' Even science studies scholar Steve Fuller admits that practitioners in the field often appear to be 'carping from the sidelines' and argues that science studies researchers should acquire at least a basic level of scientific literacy.[68]

The solution, according to Harry Collins, who has 'studied up' for decades, is to differentiate the types of competence required. Science studies researchers do not need 'procedural expertise' – the ability to do the science – but they must develop 'interactional expertise' – an ability to talk knowledgeably to experts in the field. As with any language, the best way to learn is to hear it spoken and to immerse oneself in the culture.[69]

Summary: Understanding the Divides

This study rests on the tension between two cross-cutting dimensions: public/private and basic/applied; it pays particular attention to the separating '/' . This '/' represents the overlapping interstitial spaces in which the 'open science' model and the 'network model' offer their competing explanations. The open science model was the dominant policy regime of the postwar years. It enacted a social contract for science and a linear system of innovation that justified unfettered government funding for basic science. The network model captures the zeitgeist of 'neoliberal science.' Examples include Mode 2 and Triple Helix formulations. In the strategic science policy regime, governments are more interested in funding research with direct application than in funding basic science. They deploy a dual rhetoric of research *excellence* and commercial *relevance*. Funding is contingent on cross-sectoral partnerships, market applications, and the formation of research networks. Actor-network theory (ANT) offers ways to understand the complex interactions that take place in the network forms of scientific organization that emerge under this regime.

2 Science Policy in Canada and the NCE Experiment

Three mutually interacting influences shape and constrain policy formation: powerful ideas, powerful institutions, and powerful interests act as gatekeepers to the process of agenda setting. The role of these three structuring influences in the historical development of Canadian science policy and public science institutions is described in the first half of this chapter. The remainder of the chapter focuses on the formulation and implementation of the Networks of Centres of Excellence (NCE) program as an instrument of a strategic science policy.

Historical Influences on Canadian Science Policy

Historian Donald Phillipson suggests that the Canadian state has had an abiding interest in the economic relevance of science and in promoting public- and private-sector interactions.[1] He suggests three principal reasons why this might be the case: networks of influence, fluid institutional boundaries, and international influences.

First, consistent with interest-based explanations, until quite recently 'everybody knew everyone else and everybody that mattered' at the senior levels of industrial, academic, and government science. Until the 1960s, science in Canada was very much the enterprise of a small elite group of men from similar socio-economic backgrounds who held interlocking positions of power. Their networks of influence went 'up' to the politicians, 'down' to the top Canadian talent in their own fields, and 'sideways' to senior scientists in other fields.[2]

This hegemony is illustrated in C.J. Mackenzie's response to a journalist on whether it was difficult to get government approval when the National Research Council (NRC) established a nuclear research unit during the Second World War. Mackenzie, then president of the NRC, replied,

> It was surprisingly easy. In those days the NRC reported to C.D. Howe [then minister of trade and commerce] ... C.D. was a particular friend of mine ... We all went to C.D.'s office and discussed the idea with him. I remember he sat there and listened to the whole thing, then he turned to me and said: 'What do you think?' I told him I thought it was a sound idea, then he nodded a couple of times and said: 'Okay, let's go.'[3]

For most of Canada's history, policy making was personalist: it operated on social capital rather than academic or scientific capital, and decisions were made on the basis of whom one knew. So the story of Canadian science policy is in large part the story of the people who made it. The evolution of policy attitudes towards the respective roles of basic and applied science reflects the evolution in elite ways of thinking on the topic. Although the influence of elite interests has become more subtle in recent years, it remains a major factor: 'This is Canada. When these people speak others listen.'[4]

A second element identified by Phillipson relates to institutions. Boundaries between public and private in Canadian science are quite unstable and tend to evolve fairly quickly in institutional terms. Phillipson provides the example of the Ontario Research Foundation (ORF). Founded by the province in the Depression era as a rival to the federal NRC, the ORF was transformed into a successful, autonomous, public industrial laboratory, a crown agency, in the 1950s. Later, it was 'privatized' as a state-owned corporation. Subsequently, the shares were bought by a commercial company. Another example is the Canadian Standards Association (CSA). Founded in the early 1920s as a government-funded advisory committee of researchers and industrialists, it was incorporated as a company in 1940, with the approval of a government preoccupied with war research. CSA then moved its laboratories from Ottawa to Toronto. There it became a self-financing, independent institution, and it is still authorized to promulgate and enforce standards.[5]

A third element is 'ideas based.' Awareness of other national models – predominantly American and British – has always shaped what was implemented in Canada, whether in the early twentieth century or the early twenty-first. In comparison with other advanced nations, Canadians tend to feel that they lag scientifically, and this 'national inferiority complex' has always influenced the projects undertaken. The influence of policy 'fashions' from international forums such as the OECD and the G8 can be clearly discerned in the formation of Canadian policy. Canada's National Research Council, for example, founded in 1916, was an example of convergence with similar bodies in Britain and the United States.

Taken together, the interests of powerful elites and the trade in international ideas tend to promote convergence around generalized policy regimes. However, the historical particularities of a nation's institutional and cultural legacies represent a countervailing force for divergence. In other words, Canadians put their own stamp on what they adopt. The NCE program is an example. While the phrase 'centres of excellence' was appearing with increasing regularity in the international policy discourse at the time, networking centres of excellence together was a specific solution to the peculiarities of Canadian geography (sheer size and diversity) and 'soft federalism' (powerful provinces and the requirement to serve all regions equally).

Canada's constitutional arrangements represent a long-standing constraint on federal science policy. Universities fall under provincial jurisdiction and thus are beyond direct federal reach. Historically, federal control of research funding emerged as one of the few avenues for shaping the 'national' role of universities within the 'knowledge-production system.' But until at least the 1960s, universities were not major players in the research economy. The majority of public science – historically defined in terms of utility and industrial relevance – was conducted by the NRC.

Public Science in Canada

From its inception in 1916, the NRC's 'public' mission was to serve 'private' needs by directing its research towards 'the most practical and pressing problems indicated by industrial necessities.' The obligation to serve industry was 'graven in stone' – in fact, cast on a bronze plaque beside the doors of the laboratories on Sussex Drive in Ottawa: 'This building was constructed for the purposes of fostering the scientific development of Canadian industry for Canadian needs and for the extension and expansion of Canadian trade at home and abroad.' Thus, public science was defined not as the search for knowledge but as the search for solutions.

As one of its first tasks, the NRC set out to gauge the state of industrial research in Canada. Only 37 of the 2,800 firms responding to a survey performed research on an ongoing basis, and most of these employed only one researcher. Thus there was little for the NRC to coordinate and a clear national need to develop a critical mass of researchers. This conclusion motivated the NRC to fund postgraduate scholarships in the sciences at selected universities, starting in 1917.[6]

Shortly after, the idea of constructing institutes for industrial research on university campuses began to circulate. But this heresy was briskly disposed of when proponents discovered that university faculty were adamantly opposed to 'bargaining with manufacturers.'[7] Canadian universities modelled themselves on the humanistic traditions of 'Oxbridge,' where the focus was on scholarship and teaching. To undertake research was unusual; to undertake research for industry unthinkable. The NRC's views were much the same, arguing that universities would subvert their role by conducting industrial research. The NRC itself became increasingly drawn to fundamental enquiry, if only to retain its researchers.

Between 1916 and 1940, the NRC's workforce expanded from one employee to 2,000, its annual budget from $91,600 to almost $7 million.[8] The NRC's wartime expansion allowed Canada's academic scientists to work closely with British and American colleagues on the front lines of basic advances in knowledge of microwave techniques, jet engines, digital computers, and nuclear power. Although these scientists were intent on continuing this momentum into the postwar era, conducting research within Canadian universities was still a 'fringe' activity. For example, C.D. Howe's Department of Reconstruction began an annual inventory of university research in 1946 but abandoned the project in 1949. Some scientists 'fudged figures' to conceal from university authorities how much they were diverting from teaching to spend on research. Universities were preoccupied with educating returning war veterans and other undergraduates. Research was not a priority.

But by then, the linear model of innovation was beginning to circulate as an 'international idea.' In 1951, the Massey Commission articulated the model's pipeline metaphor in noting the importance of fundamental research in priming the pump that eventually produces industrial products and applications. 'Without fundamental research,' said the commissioners, 'there can be no proper teaching of science, no scientific workers and no applied science.'[9] In the commissioners' view, basic research was most properly housed in universities, which should be adequately funded to conduct it. The commissioners strenuously opposed the idea that publicly funded laboratories should undertake research for industry, fearing that it would deaden the scientific imagination and stall the advancement of knowledge.

> [A]pplied research ... cannot be expected to add in any way to the knowledge of scientific principles. Occasionally private donors offering research grants require that research projects be approved by them. University authorities generally agree with scientists that these gifts should be steadily refused.[10]

From 1952 on, when Dr E.W.R. Steacie took the helm of the NRC, support of basic research in universities became a key Canadian policy goal. In line with the logic of the linear model, funding university research was seen as the best way for the NRC to achieve its long-term mandate to serve industry. As Steacie said, 'It is absolutely impossible to have first-rate industrial research without first-rate university research.'[11] As in the United States, the 1957 '*Sputnik* shock' had a salutary effect on research funding, helping to cement the state's commitment to basic science. Federal expenditures devoted to R&D grew from an estimated $5 million in 1939 to more than $200 million in 1959.[12]

But the policy climate began to change in the decade following the Massey Commission's report. A speculative paper submitted in 1957 by Walter Gordon's Royal Commission on Canada's Economic Prospects envisaged the roles that science might assume in the distant future, setting the stage for more intense debate on the status of science in national progress and economic development. In 1962, having examined the federally funded research system, the Royal Commission on Government Organization, chaired by J.G. Glassco, concluded that the system had failed. Glassco singled out the NRC for blame, arguing that its (vested) interests in basic, 'public' research had been promoted at the expense of applied, 'private' research:

> One of the original purposes of government in devoting money to research was to encourage and stimulate Canadian industry. From being a primary goal this has, over the years, been relegated to being little more than a minor distraction ... At present there is a wide-spread feeling that fundamental research is the only activity adequately recognized within the National Research Council.[13]

In short, Glassco famously concluded that the NRC had 'turned away' from industry. According to a funding distribution in the late 1960s, 91 per cent of the NRC budget was allocated to university research and its own laboratories (50 per cent and 41 per cent respectively), whereas only 9 per cent was allocated to industrial support and information services (5 per cent and 4 per cent respectively).[14] Commenting on reactions from NRC's scientists and bureaucrats, the OECD noted that 'many, no doubt, recognized that there were grounds for the criticism expressed by the Commission, but the majority protested against its recommendations.'[15]

Following Glassco's recommendations, a Science Secretariat was established in 1964, and the Science Council of Canada began operations in 1966. Overall, however, the Glassco framework was fundamentally undermined by a report to Prime Minister Lester Pearson by C.J. Mackenzie, the former NRC president, who advised against the substance of the findings. The personalist system protected its own. Nevertheless, the Glassco Commission report established a policy climate more hospitable to the applied/private side of the matrix. A number of government initiatives intended to bring academic research closer to the needs of industry were designed in the 1960s.[16] By the end of the decade, the Glassco Commission's main criticisms were echoed in several other policy documents, including the Science Council of Canada's 1968 report, *Towards a National Science Policy for Canada,* and an extensive survey of Canada's science and technology infrastructure by OECD examiners. The OECD and Science Council reports substantially contributed to the decade-long deliberations of the Senate's Special Committee on Science Policy, chaired by economist Maurice Lamontagne from 1968 to 1977.

Lamontagne's committee provided an exhaustive analysis of Canada's overall R&D system, the role and performance of federally funded science wherever it occurred, and the culture of science in Canada. At the core of its findings was an attack on the scientific elitism that had driven Canadian science policy since 1916. E.W.R. Steacie's proud comment that Canada stood out among the nations by recognizing 'the fundamental fact that the control of a scientific organization must be in the hands of scientists' became an indictment.[17] Such freedom, the committee argued, 'cannot be justified as a general principle for the organization of scientific progress when the tremendous cost of research has to be met mainly by public funds and when the good and bad effects of science and technology on society are becoming so far-reaching.'[18] Steps needed to be taken to bridge the gap between science and industry, and federal funding should affirm and reflect the priority of applied research.

Lamontagne was enthusiastic about the whole business of *planification* – economic forecasting and planning – and its potential for fostering *innovation*. The latter word entered the Canadian policy discourse about halfway through the 'Lamontagne decade.' Seduced by this emerging 'international idea,' the committee also embraced the new quasi-economic discipline of science policy that went along with it. Committee members and staff were thus naively enthusiastic about both the formalism of the science policy model and its appeal to actual politicians. In politics, however, extensive data can be superfluous because decision making is often a non-rational process. Politicians do not wish to be confused by too many facts: Charles Lindblom says that they 'muddle through' with small changes to the status quo, using what Michael Cohen and colleagues call 'garbage can' models of rationality.[19] According to these models, decisions do not follow an orderly or rational process. Rather, they are the outcome of the random intersection of motivated participants with previously discarded problems and solutions. Consequently, despite the years of effort that went into it, the Lamontagne Committee report, too, 'fell dead from the press,' failing to find a place on the agenda of the Trudeau administration.[20]

The power of entrenched elites to resist unwanted change is formidable, but so is the power of new elites to advance change, once the correct tools are in hand. Many politicians and bureaucrats remained convinced that the role of public science was to foster industrial innovation and economic expansion and that the NRC, with its focus on the advancement of knowledge, represented an impediment to that enterprise. As a crown corporation, however, the NRC was beyond direct political and bureaucratic interference. The only way to control it was to systematically strip away its budgets and responsibilities and transfer them to another, more subordinate, agency. In 1971, a Ministry of State for Science and Technology (MOSST) was created (as both Glassco and Lamontagne had recommended), replacing the existing Science Secretariat. In mid-decade, a 'contracting-out' policy was introduced, stating that, wherever feasible, research should be conducted in the private sector rather than in federal laboratories. In 1977, the NRC's responsibility for supporting university research was devolved to a new agency, the Natural Sciences and Engineering Research Council of Canada (NSERC), which then fell under the administrative authority of MOSST. Before long, NSERC was pioneering university-industry research programs. In 1978, MOSST also assumed authority over the Social Sciences and Humanities Research Council of Canada (SSHRC) after the Canada Council was reorganized.

Science and technology policy edged gradually towards the top of the political agenda. The first G7 summit meeting, in 1982, revealed that Canada had the lowest R&D investment in the group.[21] A Scientific Research Tax Credit was introduced to stimulate investment. Although it was a flawed instrument, open to abuse, and required a number of revisions to correct the deficiencies, it marked a major policy innovation. As a result of the changes introduced then, Canada established – and still boasts – the most generous R&D investment and tax climate in (what are now) the G8 nations. Around the same time, through the Industrial Research Assistance Agency, MOSST introduced funding to establish technology transfer/academy-industry liaison offices (TTOs/ILOs) on university campuses.

As the Liberal party came to the end of its long postwar mandate, several reports established the need to tie government support of public research to commercial relevance. In 1984, with the election of a Progressive Conservative government, the momentum towards a national science policy accelerated, and the neoliberal agenda came into play. After a period of intensive federal-provincial consultation, a national science and technology policy was formally signed in March 1987. Details of 'InnovAction: The Canadian Strategy for Science and Technology' – a $1.5 billion 'package' – were announced the following month. MOSST would be subsumed into a new 'superministry' to be known as Industry, Science and Technology Canada (ISTC), a combination that clearly signalled the alignment of science and commerce. Legislation would provide $240 million for a new 'flagship' strategy: the Networks of Centres of Excellence (NCE) program.

The Evolution of the NCE Program

The NCE program is an example of the way international ideas, existing institutions, and socio-economic interests interact under a policy regime of strategic science.[22] The policy innovation in the NCE program was to bring ideological concerns for commercial *relevance* and research *excellence* together with the concept of distributed research *networks* to form *networks* of *centres of excellence.* 'Networks' are now so associated with computers that it is hard to remember this was not always the case. By way of policy studies and science studies, the network concept was in the 1980s becoming a 'fashionable idea' in its own right, as a way of thinking about the organization of science.

The decision to embark on the NCE program was made in an ideological climate that promoted the outright privatization of public-sector functions. Where this was not possible or desirable, public-private partnerships were preferable to maintaining public-sector monopolies. Most new initiatives in science and technology partnerships saw their beginnings at this time. Canada's provincial and federal governments launched more than 100 new intersectoral research partnerships during this period.[23]

At the provincial level, Quebec's Programmme d'actions structurantes started in 1984–5 with forty networks of public-sector laboratories. Ontario's Centres of Excellence were established in 1986–7. Also in 1987, Quebec pioneered the Centre d'initiative technologique de Montréal (CITEC) at McGill University. At the federal level, Industry, Science and Technology Canada (ISTC, later Industry Canada) emphasized public-private partnerships and collaborations. Both the Natural Science and Engineering Research Council and the Medical Research Council (NSERC and MRC) actively supported collaborative targeted research. NSERC started to fund 'big science' networks in the early 1980s – in the earth sciences (Lithoprobe) and integrated circuit design (Canadian Microelectronics Corporation). During 1987–8, the budget year prior to the establishment of the NCE program, 15 per cent of NSERC's total budget went to targeted research.[24]

In late 1987, delegates to the National Forum on Post-Secondary Education raised the idea of centres of excellence that would emphasize interdisciplinarity and involve networks of researchers representing several institutions across Canada. In 1988, the Science Council of Canada advised that prosperity depended on integrating the university with the marketplace. Reinforcing this theme, in the same year the National Advisory Board on Science and Technology (NABST) recommended that greater emphasis be given to funding pre-competitive research and university-industry research consortia. This complex of initiatives and recommendations helped provide a foundational platform for the January 1988 launch of the NCE program.

Models for the NCE Program

The NCE program was designed as a hybrid of two influential models, one governmental and associated with industry, the other non-governmental and with no industrial affiliations. The first was the NRC's Industrial Research Assistance Program (IRAP), established in 1962; the second the Canadian Institute for Advanced Research (CIAR), founded in 1981.

The IRAP dates from the time when the NRC still ran along personalist ('old boy' network) lines. The IRAP's prehistory was as the Technical Information Service (TIS), founded by C.J. Mackenzie in C.D. Howe's Department of Reconstruction in 1945 and re-energized in 1962 by a retired air marshal named Ralph McBurney. TIS gave 'knowledge subsidies' to industry in the form of technical advice. The 1962 innovation added cash subsidies as well. IRAP would give grant funding to industry for private research, in the same way that universities received grants for public research. According to Donald Phillipson, the idea of giving public money to private industry was such an extraordinary precedent that it took a year's preparation by the Advisory Panel on Scientific Policy and required Treasury Board and Cabinet approval.[25]

As well as having an innovative approach to industrial research, the IRAP program was organized as a solution to Canada's geographical challenges. Rather than hire technically trained civil servants to give hands-on advice to all sorts of different industries, in every region of the country, the IRAP created a mechanism for borrowing them. Approximately two-thirds of the IRAP's field agents were locals, coopted from industries, universities, and professional associations in the region. They were paid by their own institutions, which received salary support from IRAP to release them. According to a former IRAP director, these agents constituted a 'field army' who knew their regions, identified closely with their industrial clients, and enjoyed an enormous amount of autonomy from the Ottawa bureaucracy.[26]

These 'industrial technology advisers' were gateways in extended networks of resources and facilities. Through them, small and midsized enterprises (SMEs) had access to some 130 public and private research- and technology-based organizations that were partners in the field network. In the manner that ANT calls 'heterogeneous engineering,' industry clients, their technical problems, technology advisers, provincial labs, federal labs, industry labs, engineering prototypes, and federal money were all linked together in long-chained networks dedicated to helping Canadian SMEs innovate.[27]

The networking model that began with the IRAP was clearly focused on the technical needs of industry. In contrast, the Canadian Institute for Advanced Research (CIAR), launched some twenty years later (1981) by Dr Fraser Mustard, a distinguished medical scientist, was a networking model concentrated exclusively on fundamental enquiry. Mustard and his associates promoted the idea of focusing the basic research effort in a limited number of fields where Canada had a strategic advantage and could make an original contribution. Certainly, elevating the overall pool of knowledge would benefit industry in the long run, but no immediate applications would be forthcoming.

The CIAR was conceived as an 'institute without walls,' a network that would link together outstanding researchers in institutions across Canada. According to interviews with those involved at the start, the idea came out of a dissatisfaction with existing arrangements and a realistic sense of the way knowledge works. To deal with complicated problems, some sort of institutional structure was needed that would override disciplinary and geographical barriers to the full exchange of knowledge. As well, the geographical constraints suggested that the simplest way to try to move fields was to opt for an institutional structure that invested in people rather than research.

CIAR raised funding from federal and provincial governments and from private donations, but according to informants the funding was unencumbered and in no way strategic. CIAR's mandate was the pursuit of fundamental knowledge for its own sake, without need for 'deliverables' or industry partnerships. Industry was viewed as a user of the knowledge generated rather than a collaborative partner. Funding was used to underwrite networking interactions and to buy out researchers' time at their home universities so CIAR members could pursue research on fundamental questions. The only criterion, according to a founding member, was that 'five years from now you're going to be reviewed by an international panel who will see if you have shifted the world community on how it views that question, in terms of its understanding.'

In 1986, Mustard became co-director of the committee that was designing the main features of Ontario's Centres of Excellence program, which was launched in June 1987. According to a senior civil servant, Mustard predicted that these new research centres would draw key researchers from across the country to Ontario's universities and Ontario's centres, making it extremely difficult for universities in other provinces to retain the best researchers. As a former NCE program officer put it, 'like a vortex, all the best science would migrate to Ontario.'

Earlier in the year, Mustard and one of his associates in the CIAR, Dr Patricia Baird, had been drafted onto NABST. Not surprisingly, therefore, it was NABST that brought forward the idea of creating CIAR-like national networks in the fundamental sciences to counter the Ontario initiative. The target would be fast-moving, high-profile, competitive fields that had technological implications in the relatively short term. At that stage, direct links to industry were not part of the plan. The rationale was that effective strategic or applied research programs required a good fundamental research base.

The minister and deputy minister of Industry, Science and Technology Canada paid attention to the NABST recommendations. Clearly, the federal government needed something to balance the Ontario initiative. According to a senior bureaucrat who was active in these discussions, the idea of building 'virtual' CIAR-type networks, rather than 'fixed' Ontario-type centres, was an attractive, lower-cost alternative to creating dozens of new centres around the country. The question regarding the relative merits of 'fixed' and 'distributed' centres originated in the postwar Kilgore-Bush debate regarding the creation of the U.S. National Science Foundation (see chapter 1) to promote basic research; the issue was whether the NSF should follow a 'centre of excellence' model or one that favoured a more geographical distribution of funding.[28]

Although it was interested in the network model, the ministry was not convinced that a focus on excellence in basic research was the correct route. The federal government wanted to see far more in the way of *relevance* – technology transfer to industry. The outcome was a blend of IRAP and CIAR. Like the latter, NCEs would invest in people (researchers), rather than bricks and mortar (universities and hospitals), and would be free to undertake fundamental enquiry. But, like the former, they would partner with industry and concern themselves with industry needs.

As with IRAP and CIAR, network researchers would be paid by their own institutions but would build a strong sense of belonging to a larger national entity. But in contrast to both, NCEs would be 'parasitic' on their hosts.[29] Universities and hospitals would receive no compensation for paying the salaries and benefits of network researchers, providing space and equipment, and covering laboratory overhead. NCE funds would flow to the researchers through separate network offices, which would have no duty of accountability to the university. Because their reporting allegiance was to the NCE directorate in Ottawa, these new networks would 'float' above existing institutions. They would provide the federal government with direct access to provincial university systems, overriding traditional autonomy. The networks would create a *national* research capacity open to the needs of industry and the economy.[30]

The compromise balancing 'relevance' and 'excellence' was the outcome of sustained bureaucratic struggles to capture control of the NCE initiative. The battle between the ministry and the research councils was so fierce that J.W. Pullen quickly turned it into a case study for the Canadian Centre for Management Development – the federal civil service training institute.[31]

Territorial Struggles and Program Design

Although the federal bureaucracy had been awash in rumours that a major reform of research funding was being planned, the prime minister's announcement in January 1988 came out of the blue and without any consultation with the three granting councils responsible for university research. The research council presidents quickly forged an alliance to prevent the NCE initiative being implemented without their input. The president of NSERC assigned two staff members to observe how the Prime Minister's Office was handling the new program and instructed his staff to develop alternative plans. A senior NSERC administrator interviewed the consultant hired to develop the program and concluded that the objectives would be impossible to implement (too many criteria, often conflicting). The councils discovered, as well, that public servants were to review the research applications, with final decisions made by the ministry; no peer review would be built into the process.

This contravention of scientific norms became the councils' point of attack. They argued that peer-reviewed competitions were essential to the program's academic credibility. They insisted that the councils were the only bodies with the expertise to run such competitions and to administer the resulting research funding. Without their endorsement and involvement, they suggested, the NCE program would receive a chilly reception in the academic community. If the government wanted the program to succeed, the ministry could not be allowed to control the initiative.

In May 1988, a compromise was struck. The peer-review process would be deployed strategically. Cloaked in the 'objectivity' of peer review, the program could be protected from political pressures. This separation could then be used to rhetorical advantage by the government. The Prime Minister's Office announced that the three research councils would run the NCE competition and distribute the funds, while ISTC would act as the program's secretariat. The research council presidents and the deputy minister of industry formed a steering committee, while ISTC retained overall control, albeit 'from a distance.'

As a senior civil servant noted, the ministry 'holds the pen' when writing memoranda to Cabinet or making submissions to the Treasury Board and is also 'closer to the centre' than the arm's-length granting councils. Further, two of the three research councils (NSERC and SSHRC) fall within the Industry portfolio and the minister of industry's sphere of responsibility. Nevertheless, because the ministry had no experience with research management in universities, the three council presidents exercised considerable political leverage on the steering committee. They were also able to influence the direction of intellectual enquiry, identifying as targets areas where they perceived a research gap.

As a result of the compromise, the policy objective was to reshape the culture of academic science around the dual goals noted earlier: *excellence* (fundamental research) and *relevance* (utility to industry). An Advisory Committee (to which Fraser Mustard was appointed) was established in June 1988 to design and implement the program. The committee developed four selection criteria. The weighting assigned to each reflected the success of the research councils in capturing the initiative. Research excellence was weighted at 50 per cent; a coherent, focused program of research was deemed the most decisive feature. Relevance to industry was weighted at 20 per cent, as was linkages and networking. The remaining 10 per cent covered administrative and management capability. In language reminiscent of Pasteur's Quadrant, an informant from this committee explains that '[t]he strategy was to be pregnant – we needed pure, long-term applied science that was somewhat guided by the needs of industry ... Everyone was grappling with the term "pure, long-term applied science." [It] was used to walk the fine line separating science and application.'

The program attracted diverse support. On the one hand, it was sold to Cabinet as a regional economic development package. On the other hand, it was promoted to scientists as an elitist program for producing the best science. In fact, according to one senior policy adviser, it was neither, but merely a means to pull together teams of the very best researchers who, by example, would pull the rest forward. The nomenclature of 'excellence,' he said, facilitated the process 'of capturing some of the best researchers in the country [and] recruiting them as champions for change within the system.'

Yet the program was intended to reach beyond demarcations of excellence and relevance to bring in the whole concept of research management and cross-disciplinarity. As suggested earlier, program design was much influenced by the Mode 1/Mode 2 theory of knowledge production developed by Michael Gibbons and colleagues in the late 1980s and 1990s. Gibbons served as a science policy adviser to Industry Canada during this period, and sat on the NCE selection committees. According to one informant, he was their acknowledged 'guru.' Michel Callon was also involved in the early design and implementation of the program, as a member of the International Peer Review Committee. Thus, the conceptual framework for the NCE program seems to have intended a hybrid of Mode 2 and actor-network concepts.

Following the receipt of some 240 letters of intent, 158 formal applications were forwarded for assessment to an International Peer Review Committee in November 1988. Composed of first-ranked scientists, engineers, and social scientists, mostly from the United States and Europe, this committee reported to the Advisory Committee in June 1989. As previously stipulated by the research councils, the report was made public. Public disclosure gave some assurance that the decisions were made in accordance with established scientific criteria and were not politically influenced.

Sixteen applications were deemed worthy of funding, nine in the 'must be funded' category and seven in the 'recommended for funding' second tier. The Advisory Committee endorsed all nine first-tier networks but, for reasons that remain unclear, would not support two of the second-tier networks. One of these, on ageing, was the only social science proposal on the short list. According to an Advisory Committee informant, after extensive lobbying by the councils, 'a decision came from above' to include the ageing network. However, it would be funded by the research councils rather than the NCE. The poor showing of the social sciences was later attributed to selection criteria oriented towards engineering and the hard sciences rather than the broad perspective needed to make the participation of human scientists possible.

Because they reflected a compromise, the initial selection criteria failed to fully articulate the preferences of either the research councils or the ministry. In practice, networking and industrial relevance hardly figured into the equation. And because companies made few cash commitments at the proposal stage, it was difficult to assess the extent of partnerships and linkages. Academics inexperienced in such matters found it difficult to demonstrate such competencies. For similar reasons the applications were weak in defining proposed management structures. Furthermore, the reviewers themselves were not skilled in assessing this area. As a result, states a member of the Advisory Committee, the reviewers

> could not bring themselves to say no to the best science, regardless of the other criteria. They could not displace top-quality science with inferior science just because they had a better management structure or because they scored so high on practical application. The other three criteria were ephemeral, intangible, hard to measure or understand. [Reviewers] could not bring themselves to knock out top science on the basis of criteria they did not understand and could not operationalize.

In the end, the reviewers decided to gamble on the best science and hope that everything else fell into place.

Mobilizing Networks: Changing Attitudes

The NCE program introduced two radical and important hypotheses, according to Stuart Smith, chair of the International Peer Review and Implementation Committees. At a November 1989 briefing session for the winning networks, he told participants that the first hypothesis would test whether collaborative research could be done at a distance using telecommunications technologies. The second would test 'whether it was possible in the field of long-term and fundamental research to force researchers to think about the economic and social impact of their work, and more particularly about the channels by which the research results will be commercialized.'[32]

The federal bureaucracy had no operational framework for the implementation of NCE policy. Ottawa and the networks made up and modified rules and expectations as the concepts evolved. One of the tasks of the NCE directorate, in the early years, was to convince scientists that their responsibilities extended beyond the standards of traditional funding programs, and beyond the norms of academic science. Program staff realized that researchers initially viewed the program as just one more funding source for basic science. As one policy adviser noted, 'The scientists didn't know what they were getting into. They just went into it for the money. Very clearly at the start, it was just another pot of money with some arbitrary rules that they would pretend to follow.' It was necessary to convey the 'expectation that [they] were going to interact with industry and that there was going to be some kind of measurable outcome from that interaction,' according to a senior civil servant.

For the networks, that first phase was all about inventing themselves, consolidating themselves, establishing relationships among researchers, host institutions, and industry partners. Industrial partnerships were slow in coming. Sources agree that a great deal of 'courting' went on in Phase I, but not a lot of commitment. The first year, fiscal 1991, was only a partial year. Networks spent most of their time establishing the mechanics of administration – systems, committee structures, and so on. After that, only three full years remained before funding ended. At that point, no guarantees had been given that the program would be renewed. The program was experimental. As far as anyone knew, four years total was all they had.

That situation changed in December 1992, when Brian Mulroney's Progressive Conservative government brought down its final budget.[33] In the same speech that abolished the Science Council of Canada and the Economic Council of Canada, Finance Minister Don Mazankowski announced that the NCE program would be extended.[34] A new competition would be held in targeted areas, and existing networks would be able to compete for a second four-year phase of funding (fiscal years 1995–8). The decision was supported by a positive interim program evaluation carried out between July and December 1992. The evaluation reviewed the effectiveness of program and network management, the level of networking, and the nature and extent of industrial involvement. From the tenor of the announcement, it was clear that the last named was deemed less than satisfactory. In order to be renewed, networks would have to deliver much more in terms of commercial relevance and industry partnerships. As the 1992 federal budget announcement stated,

> From the beginning, the need for industry involvement and cooperation in the networks has been stressed. Given the need to strengthen this kind of industry collaboration with the research community, funding is being extended. This will ensure that the most successful of the existing networks continue to contribute to competitiveness.

A reduced budget of $197 million was allocated for the four-year period 1995–8, with 25 per cent set aside for developing the planned new networks. Modified selection criteria reflected the shift in emphasis from *excellence* to *relevance* and precipitated the dilution of meaning mentioned earlier. Now five criteria, all equally weighted, had to exceed an established 'threshold of excellence':

- excellence of the research program: 20 per cent (was 50 per cent)
- training of highly qualified personnel: 20 per cent (new)
- networking and industry partnerships: 20 per cent (same as before)
- knowledge exchange and technology exploitation: 20 per cent (new)
- network management: 20 per cent (was 10 per cent).

As a senior civil servant noted, ISTC had successfully reoriented the program to something that they were more comfortable with. The new criteria reflected what they had wanted from the start: a program that fostered more industrially relevant research. A rotation of research council presidents helped consolidate this position. The new leaders of the MRC and of NSERC were more focused on developing university-industry linkages and on having academics work outside of their traditional environments of interaction.

The attitude of faculty was more ambivalent. The top-down decision to shift priorities represented a serious concern for some of the researchers involved, and considerable turnover among scientists occurred. Some found the program more appealing and enlisted; others recognized that they would not fit and left. Since Phase II, according to a senior civil servant, all networks have conducted more applied and less fundamental research. Reduced budgets for the renewed networks forced the scientists to focus more on lines of research that were likely to be of interest to industry. The research still had basic components but was aligned to be of greater interest to the existing industrial environment.

With the election of a Liberal government in October 1993, the emphasis on relevance became even more entrenched. By now, 'neoliberal' principles had become a political orthodoxy as even centrist parties shifted to the right. Shortly after assuming office, the Liberals undertook a massive reorganization of ISTC. As if to confirm the subordination of science to the economy, the department now became simply Industry Canada. It assumed a much-enlarged portfolio and a mandate to foster Canada's international competitiveness. The following year, 1994, a major science and technology program review was announced, together with the intention of moving towards a new national science and technology strategy. Months of exhaustive consultation and review followed. After some considerable delay, the new national policy – 'Science and Technology for the New Century: A Federal Strategy' – was finally announced by Industry Canada in March 1996.

The strategy adopted science and technology as a federal priority. Taking a national system of innovation approach, it integrated academy, industry, and government research under the rubric of job creation and economic growth. The focus was on 'the strategic investment of resources for the maximum economic, social, and scientific returns.'[35] The principal means of achieving this was through the strategic use of public-private research arrangements between and among universities, industry, and other levels of government.

Both the Conservative and Liberal administrations had crafted a climate hospitable to commercial relevance by applying a multitude of mutually reinforcing policy instruments. Their efforts appear to have been successful. Intellectual property rights were becoming the default currency of the research economy. Gross revenues from royalties and licences grew more than threefold between 1991 and 1997, for example, while industrial research funding saw more than a fourfold increase.[36] Royalties more than doubled again between 1999 and 2001, from $21.1 million to $47.6 million. The rate of company creation reflected the same pace. Of the 680 spin-off companies created in Canadian academic institutions in the period from 1980 to 2001, more than 78 per cent had been formed since 1990, at an average rate of about 50 per year.[37]

Industrial support of academic research has advanced more rapidly in Canada than elsewhere in the original G7 nations. Table 2.1 shows a

TABLE 2.1
Share of academic research funded by industry, G7/G8 nations, 2000, 1997, 1990, 1985 (percentages)

	2000	1997	1990	1985
Canada	8.9	9.7	6.3	4.3
United States	6.0	6.0	4.7	3.8
Japan	2.5	2.4	2.3	1.5
France	2.7	3.0	4.9	1.9
Germany	11.6	9.7	7.8	5.9
Italy	–	–	2.4	1.5
United Kingdom	7.1	7.1	7.6	5.2

Source: OECD (1998: 165) and OECD (2002-2: table 48)

general upward trend in the proportion of industry funding for academic research between 1985 and 2000. Canada and Germany share the lead until 2000, when Canada's renewed commitment to support of the public-sector research base – a reversal of the policy of funding cuts that began in 1996 – is reflected in a slight downward correction. The significant and long-term federal reinvestment diluted industry's *proportion,* although the *dollar* figure continued to climb.

Table 2.2 shows that in 2001 Canadian universities performed a higher percentage (32.7 per cent) of national R&D than did other G7/G8 countries. However, the business sector in Canada remained a low performer (55.9 per cent) and financed a lower proportion (42 per cent) of the nation's research and development effort than in other nations, suggesting that Canada's industries continued to rely on publicly supported research rather than developing their own infrastructure. Note also that the countries most open to foreign investment were Canada and the United Kingdom, with 15.5 per cent and 16.3 per cent, respectively.

The reduction in federal funding support for science and technology that was initiated in 1996 caught the NCE program in its net. The program was slated for cancellation until the networks came together to launch a public relations and lobbying campaign. Their efforts were successful, and the NCE program was made permanent in the February 1997 budget,[38] albeit with a 'sunset clause.' The purpose of the clause, according to a senior civil servant, was to allow the program 'to continuously reinvent itself through a constant influx of new people

TABLE 2.2
R&D expenditures by financing and performing sectors, G7/G8 nations, 2001 (percentages)

	Financing sector				Performing sector			
	Industry	Foreign	State	Other	Industry	Academic	State	Other
Canada	42.0	15.5	32.1	10.3	55.9	32.7	10.6	0.9
United States	68.2	–	27.3	4.4	75.3	13.6	7.5	3.6
Japan	72.4	0.4	19.6	7.6	71.0	14.5	9.9	4.6
France	54.1	7.0	36.9	1.9	64.0	16.7	17.8	1.5
Germany	66.9	2.1	30.7	0.4	71.4	15.5	13.1	–
Italy	–	–	–	–	49.3	31.5	19.2	–
United Kingdom	49.3	16.3	28.9	5.5	65.7	20.7	12.2	1.4

Source: OECD (2002-2: tables 9, 11)

and ideas.' The networks least likely to survive without government support would be culled, funding to those deemed to have 'graduated' from the program would be discontinued, and funding for all networks would be capped at a maximum of fourteen years. For the surviving original networks therefore, Phase III would be the end of the line. Policy makers did not intend NCEs to become entrenched and institutionalized. They wanted researchers to understand, from the start, that NCE funding was finite.

Many viewed the cap as arbitrary, given that funding was allocated on a *competitive* basis, to the best proposals. Why should they not continue to compete and be judged on their merits? The change created detrimental amounts of goal displacement among networks in the life-sciences sector. Instead of focusing on advancing fundamental and translational research, networks facing sunset focused their attention on speculative financial projects in order to replace federal funding.

Nevertheless, the intention of the NCE initiative was, in part, to generate precisely this kind of cultural change in academic science. The program's biggest achievement, according to one program officer, has been to establish 'a market orientation in academic researchers and a predisposition for collaborating with the private sector.' This included finding and developing receptor capacity in Canadian industry, securing venture capital, negotiating multiparty intellectual property agreements, and establishing an effective process whereby network technologies could be licensed to industrial partners. As program documentation indicates, the numbers of patents filed and inventions disclosed increased significantly. Sophisticated alliances with the financial sector allowed some of the networks to attain experiential knowledge of business and finance that often surpassed that of directorate staff. They knew what was needed to run their own programs and felt constrained by the pedestrian advice of NCE officials. Not surprisingly, the networks began to take on a 'life of their own' as they claimed increasing autonomy. As one senior civil servant put it,

> We started to see change where the people who were working in the program had a very strong concept of what it was that they were doing. It wasn't always exactly the same as our concept, but they began to drive the program in certain ways ... We [government] still set the agenda, but the level of contribution is much higher from the networks now, and I

> would say that many times now we are learning from them as opposed to them learning from us ... We started to see a change from us really driving the program to them taking much more ownership for it and starting to push into new directions.

The NCE directorate became somewhat uneasy with the aggressive commercial ethos that developed in some of the networks. They sensed that things had gone too far. In its review of one of the life-science networks, for example, the 1997 Phase III Selection Committee Report suggested that the goals of the research program needed redirecting in order to be more appropriate for an academic setting. In other words, the network 'should not try to compete in areas of research where major pharmaceutical companies are already investing enormous amounts of money and have a clear research lead and advantage.'[39] But these Phase III funding proposals were prepared by networks facing the sunset of NCE support. They were required to show how they would handle the transition. It was almost inevitable that they would respond in commercially aggressive ways.

In recent years, many of the networks have formally incorporated themselves to facilitate the management of their extensive research programs, intellectual property portfolios, and partnerships. Incorporation was always Industry Canada's preference. They saw formal, legal structure as a means of eliminating the model of collegial governance that had guided academic decision making in the past. But the research councils resisted, preferring to leave the decision up to the individual networks. After initially adopting a 'wait and see' position, most have now incorporated. They have also created arm's-length, for-profit corporations that use standard business tools such as mission statements and strategic plans.

A decision to incorporate raises some interesting conceptual issues. A network is a loose association of researchers, nodes, projects, and partners. It *is* the people and entities that make it up. But a corporate body has legal powers of association and personhood. It exists *apart from* the people and entities that make it up. An incorporated (literally: *embodied*) network seems almost contradictory. Incorporation institutionalizes these 'virtual' entities, cloaking them in substantive legality. The increasing adoption of the corporate form signals the approaching funding sunset for 'mature' networks and their desire to sustain themselves beyond this horizon.

Summary: The Implications of Policy Change

In Canada, as elsewhere, national policies promote the integration of public-sector research organizations into the economic mainstream: public science must move out of academic and government labs and into the marketplace. Policy goals include the commercialization of research results as proprietary products and the adoption of new, market-friendly institutional arrangements for the conduct of research. Policy tools such as intellectual property rights and public-private research networks promote the development of closer academy-industry relations and facilitate what can loosely be called the privatization of the public knowledge base. Yet at the same time that they promote commercial *relevance,* these policies also promote scientific *excellence* – a combination that may at first appear counter-intuitive.

But Canada has a long tradition, stretching back into the nineteenth century, of state involvement in the promotion of programs that seek both relevance and excellence.[40] The National Research Council (NRC) was founded in 1916 largely to address the needs of industry for research that would advance innovation. At a time when universities were in the business of humanistic scholarship and teaching, rather than the advancement of scientific knowledge, the establishment of the NRC represented the institutionalization of federal attempts to advance 'useful' research. Over time, however, this intent was subverted as the NRC became increasingly focused on conducting fundamental research and promoting the same in universities.

Beginning in the 1960s, attempts at policy reform proposed ways to 'correct' the orientation of federally funded research and scientific cultures and to turn public research towards economic development goals. The scientific establishment successfully resisted these attempts until the 1980s, when the neoliberal turn in Canada's political culture established a strategic science regime that would harness public science to the needs of the economy. One of these initiatives established the Networks of Centre of Excellence (NCE) program.

As a hybrid of the NRC's Industrial Research Assistance Program and Dr Fraser Mustard's Canadian Institute for Advanced Research, the NCE program was dedicated to *both* scientific excellence *and* commercial relevance. Because of its novelty and dual commitments, the program was the subject of fierce jurisdictional struggles within the federal bureaucracy as the research funding councils and the ministry responsible for industrial and economic expansion fought for control. In the first phase of the program (in the early 1990s), the culture of the research councils dominated, and scientific excellence was the primary selection criterion. This phase was concerned more with basic research performed under 'open science' conditions in public institutions, or 'Bohr's Quadrant.' In the second phase (in the mid-1990s), Industry Canada's concerns for commercial relevance came to the fore. As a result, some of the networks entered into market relations more aggressively than had been anticipated, in pursuit of applied research for private profit, and moved into 'Edison's Quadrant.' After 1997, when the program became permanent, new networks were selected on more balanced criteria and relevance was redefined in social as well as economic terms. 'Pasteur's Quadrant' was the goal.

But this goal has been pursued throughout the program's history. Mechanisms have been sought that will couple the creation of knowledge and traditional means of diffusion, such as journal articles, with 'translation' of knowledge and new means of diffusion such as technology transfer to industry partners. The two are rife with tension, and ways have been sought to reconcile, for example, publication norms with the protection of intellectual property rights – or, when considering whom to recruit into a network, to reconcile traditional criteria of scientific merit with strategic judgments of a research program's commercial relevance. Control of these and other tensions is accomplished within the formal organizational and management structures the program requires networks to adopt.

It is possible to develop a model of Canada's strategic science policy regime, and the way the NCE program relates to it (see figure 2.1). This model, with modifications for local conditions, may be generalizable to other countries operating under a similar regime. It shows the influence of powerful interests, ideas, and institutions at the agenda-setting stage of policy formation. Once an agenda for commercial *relevance* and scientific *excellence* is mobilized, competing state agencies (in this case, Industry Canada – IC; and the research councils – RC) place countervailing pressures on the research culture and attempt to influence the development of policy instruments that will further their interests. The NCE program is such an instrument. The construction of 'networks of centres of excellence' is intended to promote the formation of human, social, and economic capital, leading to a new national capacity in research, improved receptor capability in industry, and a new research culture. These results would then legitimate

Figure 2.1: Canada's strategic science policy regime in relation to the NCE program

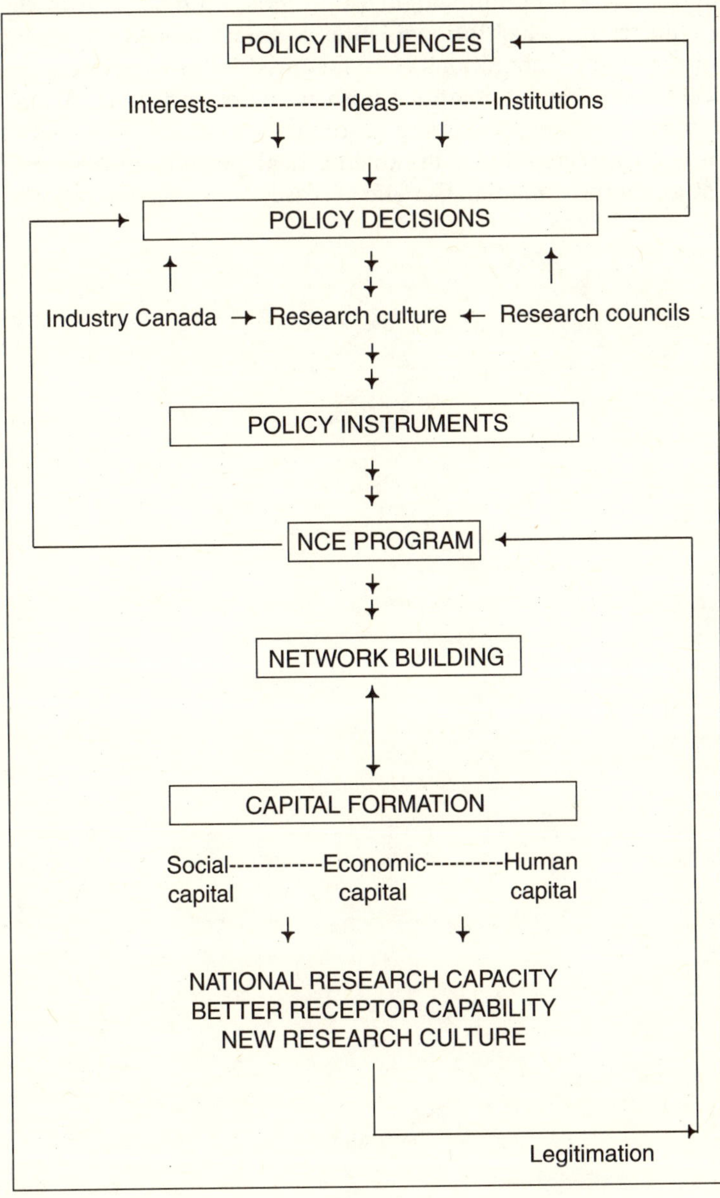

such programs and encourage the development of other similar initiatives.

Overall, the NCE program sought to promote a broad shift in the research culture. Inter-institutional, inter-sectoral, cross-disciplinary, and multi-regional collaborations were favoured in the network selection process. Constructive relations with industry and cost-efficient, even revenue-generating, operations were to be pursued. The extent to which these goals were achieved is an empirical question addressed in the case study of the Canadian Genetic Diseases Network that follows.

3 Configuring the Canadian Genetic Diseases Network

What is a network? Often, it is thought to be something flimsy or ephemeral, like a cobweb, that can easily tear and drift apart – just a web of relationships, with nothing visible anchoring them in place. But this is not the case. Networks are anchored in the materiality of the actors that make them up: in the infrastructures actors inhabit, in the resources actors command, in the allies they enrol, and in the artifacts and instruments they employ (or, as is often the case, are employed by). Thus actors both 'come before' and 'make up' networks in a complex process of emergence and consolidation. This chapter is about that process and how it is enacted in the Canadian Genetic Diseases Network (CGDN).

The first part of this chapter examines how the CGDN knitted the first few stitches of a previously non-existent web and how it secured itself to the material foundations of universities. It outlines the vision of the network's scientific director and the peer group he enrolled at the start. Flowing from that, chronologically, is a description of the network's genesis in 1988 and the recruitment of the founding researchers and professional staff. This discussion ends with a description of the management structure and the formation of an institutional identity that is separate from the university. The second section is a critical analysis of problems that have emerged from the way the network has been configured. The problems are associated with a variety of issues: regional distribution, elitism and equity, social reflexivity, and public accountability.

The Power of One

To succeed within the parameters stipulated by the NCE program, member networks seem to require strong, even visionary scientific leaders, people who perceive the program as a means to animate and execute their vision. In actor-network theory, power flowing through networks eventually accumulates in the hands of the most influential actors: those able to identify and enrol more potential allies than their rivals, align[1] the interests of these allies with those of the network, then act as spokesperson for the network as a whole. Rather than being delegated by pre-existing groups to speak on their behalf, *spokespersons* actually create the groups they speak for, simply by assuming that role.[2] This ability to intervene consolidates power and agency over time.

Strong Leadership

Generative leadership was common to all the networks created in Phase I of the NCE program, but particularly those in the life sciences. Strong leadership is consistent with the culture of molecular biology, where the laboratory leader focuses all the resources and recognition of the lab, and represents the entity as a whole to the lab's various communities. The leader functions 'as a symbol of the lab as the lab's information interface, its "provider," and as the one who plays the games of the field.'[3] In CGDN, that spokesperson was its scientific director, Michael Hayden. His vision, communicated in a January 1991 essay entitled 'Science and Dreams,' was 'to create a functionally integrated but spatially dispersed intellectual consortium ... to open new pathways for collaboration and networking while breaking down the old style, conventional, departmental and institutional barriers. This is not business as usual.'[4]

All interviewees agreed that Hayden was the person most responsible for the network's initial success and that he remains its biggest influence. In many cases the opinion was volunteered rather than solicited. He conceived the network, envisaged its framework, and personally enrolled most of the researchers and staff. He is often characterized as 'a network in himself': his contacts and force of personality stamp the network's style as entrepreneurial and fast moving. According to one former NCE program officer, 'Hayden is unrivalled as a scientific leader. He was the right person in the right place. He was certainly the most effective of the scientific leaders I observed.' Hayden appears to command the loyalty and respect, even the affection, of his colleagues. Typical expressions of support were expressed by two researchers and a network executive:

RESEARCHER: He has actually made this one of the most, if not the most, successful networks out of all those centres of excellence that were set up.

RESEARCHER: It's very strongly led by Michael Hayden. He has maintained the leadership through the whole time. He's certainly done an excellent job. I think it's very much his baby.

EXECUTIVE: Simply put, Michael Hayden is a wonderful, wonderful, network leader. He always has been, right from the beginning. He's a rare combination – a person that's guided by principle but tremendously goal oriented. He knows what he wants to accomplish and he is tenacious. He won't let go of an objective he believes in, and he believes in the network.

Hayden's leadership style is characteristic of the traditional command-and-control (Mode 1) model of academic science, in which senior scientists exercise almost total control of their eponymously named laboratories. This is the milieu in which the current generation of researchers was socialized. It is not surprising, therefore, that Hayden runs the network, in the words of a recent recruit, as 'a benevolent dictatorship,' or that everybody seems to accept autocracy as the natural order: '[T]his is not a democracy; one cannot run a network like this like a democracy. Michael Hayden makes most of the decisions. He has the best background. He's the best choice. So it runs quite smoothly.'

An external observer noted that Hayden provides strong scientific leadership, but that his style is less collaborative and consultative than some. 'Hayden sets scientific directions by force of personality, although he seems to do so without ruffling too many feathers. Not necessarily bad, but different from the other two [Phase I, life sciences] networks I think.' One of the NCE program officers – all of whom are scientists themselves – explained it this way:

> It's not really a dictatorship. You have to understand the scientific community that you're dealing with ... It's a highly educated population. A highly critical, opinionated population. We are trained to be very critical of each other's work. So when you're dealing with that sort of culture it requires very strong leadership. Others might equate it to dictatorship but it is not. You have to be able to stand strong against all of the criticism. And so the leaders have to be very strong. And very firm. Because it's not going to work otherwise.

Hayden's way, explained a senior researcher, is to put his imprint on something and set the strategic direction, then hand it over to professional staff and move on to something else. 'He has the final word, but those people are now so indoctrinated that they run on their own. They don't need to go to him for everything. And it works.' A veteran staff member agreed. Hayden makes the decisions and sets direction, she said, but, over the years, 'he backed off and let us do our own thing.' A senior science bureaucrat who was the network's program officer for a number of years noted that Hayden indeed did less hands-on management than most of the other leaders. 'But when he did intervene,' she said, 'he had vision and a pretty good schtick. He really got things done.'

Hayden's willingness to allow the network's professional staff to manage network affairs was, in part, an artifact of the program's design. As described later in this chapter, a major novelty of the NCE program was that network research was conceived as managed research. Given the large amount of funding allocated to each network and the complexity of linking so many institutions and researchers together, formal management structures were deemed essential. In effect, each network had *two* leaders. One was the scientific director. The other was a network manager, who, according to a policy adviser, 'made bloody sure they knew what everybody was doing. And kept tabs on everything. Which is very unusual in a science program.'

As scientific director, Hayden coordinated and integrated all the research projects and programs. But the network's senior executive officer controlled the spending and monitored the researchers to ensure that all the network's non-scientific mandate points were being met. In accepting the position, noted one of these senior staff members, he knew that working alongside Hayden would be demanding but felt confident enough to accept the challenge. 'I could work with him long enough to work it out. You just have to be strong. He backed me and I backed him; it worked both ways.'

Part of Hayden's success as a leader came from his strategic abilities. He was able to mobilize resources, at the last minute, for the highest impact. For example, the face-to-face aspects of funding applications – expert panel visits, presentations to the NCE selection committee, and so on – were orchestrated to maximum effect. According to informants, every ally, every board member, every industry partner, every network scientist was invited to sit at the table. Everyone gave five-minute presentations on their research and/or role in the network, literally overwhelming panelists with information and enthusiasm for the science.

These funding reviews and site visits were highly polished performances. Everyone was well prepared. The whole effort was timed and scripted, without appearing slick. As the former managing director described it, 'Everybody was there to back up that this organization was doing its stuff ... you can't leave anything to chance, you have to cover all of the bases.' Hayden himself, however, relied on staff to set things up, rarely focusing until the very last minute. He caused more than a few anxious moments, but people learned to have faith in his ability to deliver the goods. The following anecdote, told by an NCE program officer, provides an example of Hayden's eleventh-hour style.

> I never saw anybody like Michael for pulling things off at the last minute. I'd talk to him one day and he'd have to do something the next day and he would be totally disorganized. And I'd expect an utter disaster. And then the next day I'd see him perform and he always seemed to pull the rabbit out of the hat. Yeah, the lights went on and Mike was there. He'd just put in a terrific performance and really inspire people in the network.
>
> The night before the selection committee meeting [for Phase II] there was a dinner for Michael Smith in recognition of the Nobel Prize. And Michael Hayden was at the dinner and I talked to him and he was really nervous about appearing before the selection committee the next day and all that went along with it. And I thought, 'Oh God! He is unprepared. He is going to bomb,' you know?
>
> But when he came in the next day he did a really smart thing. He brought in a private-sector partner to say what was great about this network from industry's point of view. Hayden was the only person who did that. Everybody else brought in their scientific director and their management person. So his network was unique in that way. And that was exactly the dimension that the committee wanted to hear.
>
> A distinguished member of the selection committee ... quite an influential guy, said 'You know we can't *not* fund this guy. This guy shakes trees.' And I always remember that and it certainly is true. Michael really did have that impact.

Involving so many network members – scientists, board members, and industry partners – in the renewal effort was extremely innovative at the time. Not all networks took such an inclusive approach. For example, a researcher from another life science network – who also belonged to CGDN – reported having few companions when he attended a renewal panel. The leaders had invited only three or four scientists to present a synopsis of what was happening in that network. 'None of the other scientists was invited; it was only a handful of people.' That network subsequently lost funding largely because, according to this researcher, they had failed to engage their scientists in the process. Unlike Michael Hayden, that network's leadership 'essentially excluded all the scientists and then tried to move forward. But of course, they had nothing left. The scientists had abandoned ship.' At CGDN, in contrast, 'every one of our scientists was at the review committee meetings. No one was missing unless their mother was dying. There were no excuses. You had to be there.'

Thus, the essence of Hayden's scientific leadership was to involve others. Hayden extended that concept of involvement to the wider community. He invoked the notion of 'civic science,'[5] arguing that, when scientists accept public money, they also accept a responsibility to the communities that provide those funds. Science and scientists must not be cloistered, he said; they must participate actively in society and be fully accountable. The obligation is not so much to the government as to the public at large. In return for the privilege of being funded to practise science, scientists must accept the responsibility of ensuring that the community understands what they do. Hayden claimed that facilitating this understanding was as important as his work on human health. 'We have a responsibility to reach out to the people who support us ... We are *guests* of the public. And so we have a responsibility to acknowledge that they are the source of what we're doing, and why we're doing it.'

Although 'civic science' sounds high-minded, it is more about furthering public *funding* of science than about promoting public *understanding* of science. To use the vocabulary of ANT, when scientists are astute about *enrolling* and *mobilizing* the public as *allies* – when they convey a convincing message – the public will pressure politicians to maintain or increase funding levels. According to Hayden, the cuts to the basic research budget in the mid-1990s occurred because scientists 'were not civic enough. And so people didn't place enough priority on it.' Seeing what was happening to other programs, NCEs 'had to get out there and make sure [network] research was high up on the political agenda. Governments *do* respond to the people, particularly around election time.'

The reference here is to the federal program review of 1996. As part of deficit reduction, the federal government had decided to discontinue NCEs. The winding-up process had begun; no more funding would be forthcoming. In response, the networks, led by CGDN, waged a national public relations campaign to save the program. As a senior network manager explained, 'it took about four months but we won. We won big ... We convinced the government that this was a program that they couldn't afford to let die.' In other words, through ANT's process of *interessement* government had been persuaded to define its problem in such a way that the NCE program was the solution, the *obligatory passage point* for strategic science.

Since then, according to Hayden, network scientists have been 'tremendously civic.' In every part of the country, 'they are out there talking to the wider community.' Perhaps partly as a result of the mobilization of public sentiment in this way, scientific research recovered its place on the policy agenda. As the deficits turned into surpluses, former funding levels began to be restored, then equalled, then exceeded. (For example, within three years of the 1998 founding of the Canadian Institutes for Health Research, its budget was twice that of the Medical Research Council it had replaced.) Research funding was back on the federal 'radar screen' as a priority item in the federal budget for four consecutive years (1998–2001). Powerful 'advocacy coalitions' mobilized to lobby for NCEs.[6] Program funding almost doubled between 1997 and 1999, from approximately $40 million a year at the end of Phase II to $78 million a year in the 1999 and subsequent budgets.

Civic science can thus be seen as a rhetorical strategy that aligns scientists' self-interest with the public interest by enrolling the public as allies in the network. 'By doing it,' said Hayden, 'we ensure our future.' While mobilizing public support for science funding is a legitimate activity, some observers find something slightly 'slick' about the way Hayden packages it. One critic, a senior scientist and policy consultant, said, 'Mike Hayden is what I would call an *operator*. I do not mean this in a terribly critical way. It is just the sort of person that he is.' Another senior scientist criticized Hayden's ability to present genetics as the solution to a host of medical problems, thereby diverting attention from the complex 'web of causation' in disease of which genetics is but a minor part.

Bruno Latour has shown that Louis Pasteur *the scientist*, who made fundamental discoveries in microbiology and public health, is inseparable from Louis Pasteur *the politician*, who skilfully mobilized legions of microbes, farmers, laboratories, and other allies to create new sources of social power and legitimacy. Pasteur became *Pasteur,* the authorized spokesman and exclusive interpreter for the heterogeneous multitudes he enrolled. Hayden had extended his personal network and entrenched his leadership role over the decade of the network's existence. Like 'Pasteur' in Latour's analysis, 'Hayden' had become the authorized spokesperson for legions of molecules, machines, and tests; patients, doctors, and researchers; founder populations; government funders; disease foundations; and pharmaceutical interests.[7] By interesting and enrolling powerful allies and mobilizing the rhetoric of medical genetics in the public arena, Hayden's science had become a political practice, a science of *associations,* what ANT calls 'politics by other means.'

To understand 'Hayden' and 'CGDN' as consolidated complexes of linkages, it is helpful to map the beginnings of the network, before any taken-for-granted relationships were stabilized. In the early days, Hayden reached out to senior colleagues to help build the network. He was enrolling a core-set: a nucleus of allies comprising the best in the field of medical genetics in Canada.

Enrolling the Core-Set

In January 1988, when Prime Minister Brian Mulroney announced funding for a new initiative called the NCE program, Hayden, then a young associate professor at the University of British Columbia, immediately saw the potential for a genetics network. As a relatively junior researcher, however, he needed to enrol established members of the genetics community for a proposal to succeed. He telephoned the two top medical geneticists in Canada: Charles Scriver (an expert on Tay-Sachs and PKU) at McGill University, and Ron Worton (discoverer of the Duchenne muscular dystrophy gene), then at the University of Toronto's Hospital for Sick Children ('Sick Kids'). Hayden had not worked with them previously, but he knew their work and their stature and, in effect, told them, 'You know, we've got an opportunity here for a network in the genetic basis of human disease.' Worton had been thinking along similar lines himself and was willing to work on it with Hayden. Scriver was more circumspect. Hayden said, 'I was really young back then, and Charles was like the Father of Genetics. Why would he care? And why would he trust me enough to work with me on this?'

Scriver was an essential ally for several reasons beyond his scientific seniority. First, twenty years earlier, in 1969, he had helped found a well-known program called the Quebec Network of Genetic Medicine. That network ran a screening program for newborns and a distributed system of centres providing diagnostic follow-up, genetic counselling, and treatment. The group had recently published an article in *Science*'s first theme issue on how science could contribute to societal initiatives and concerns. Scriver suggests that within this context the network's name and structure attracted Hayden's interest. Second, Scriver had research projects funded under Quebec's Programme d'actions structurantes. That provincial program, formed in 1982, appears to have been one of the prototypes of the federal NCE program, formed in 1988. Like NCEs, actions structurantes projects had to be performed by a team of investigators. While industry partnerships were not required, teams had to be multi-university and multidisciplinary.

Scriver was coming to Vancouver the following week on a personal matter. Hayden arranged a meeting. The two researchers, separated in age by a generation or more, sat on the steps of the Vancouver Art Gallery, in the chilly middle of February, going over the issues. Hayden summarized the federal announcement and pointed out the similarities with what Scriver had built in Quebec. He talked about pulling together the 'best of the best' across the country, in the same way that Scriver had pulled together the 'best of the best' in Quebec. He pointed out that millions of dollars were being made available for research. Finally, he asked Scriver whether he would join in. Hayden's energy, passion, and rhetorical skills are hard to resist. He convinced Scriver to participate. Hayden calls it 'a pivotal conversation.' Scriver said of his recruitment, 'I think Michael recognized an interesting opportunity when he saw it, which has been his trademark all along. He was aware of what we had been doing in Quebec with bringing academic genetics to a societal interface, and he thought that would make an NCE proposal look good.'

All three had their own personal networks of colleagues and contacts and technical capacities, and these quickly combined and multiplied the way networks do, sparking from node to node. Hayden, Scriver, and Worton were thus the embodied 'centres of excellence' from which the network originally sprang. Senior members of the network soon called them 'the triumvirate.' Beyond these three founders was the elite group of scientists they enrolled to craft the initial letter of intent and subsequent proposal. Worton recruited two people from Sick Kids – Lap Chee Tsui and Rod McInnes – while Scriver brought in Roy Gravel and Emil Skamene from McGill University. Together with Hayden, that made a core-set of seven.[8]

This 'group of seven' met in a Toronto hotel room for a day and a half to brainstorm ideas. But that first session was followed by a long hiatus as they waited for the government to specify what was expected in the letters of intent. As Ron Worton recalled,

> The next thing I remember is that I'd planned a three-week holiday for that summer and I'd just bought a cottage the fall before. So this was my first summer in my new cottage. I had never had a three-week holiday before. This was going to be my first lengthy vacation. And I'd been there about a week and a half and I got a call from Michael and he said he'd just heard that NSERC – the leaders of this program at the time – were doing a cross-Canada tour talking about the network model and how to apply and so on. The tour would be in Toronto the following week.
>
> That ended my three-week holiday. I went back to Toronto, listened to the presentation and took notes and called Michael, and two weeks later I was with him in Vancouver. I guess we spent the best part of that summer putting together the letter of intent ... and then ... in the fall, it had to go very fast ... We only had six weeks between notification of the success of the letter of intent and the requirement for the proposal.

The core-set identified and enrolled people in other universities and hospitals, expanding in multiples, from the original group of seven, to fourteen, and then to twenty-one for the formal proposal. Roy Gravel remembers recruiting people into the program during the summer of 1988. 'I recall there was a meeting in Toronto, the Genetics Society or something of this sort, that was North America wide. It brought a lot of these people into the city. But that was very close to the deadline. We already had most of the people identified by that point.'

One of the most novel aspects of the NCE initiative, one that caught the attention of scientists, was that research was to be extended across Canada in east-west interactions. This was not the traditional way in which Canadian science had been organized. Few national forums brought Canadian scientists together. Most connections and collaborations were north-south: Canadian scientists tended to meet each other, if at all, at conferences in the United States. As a result, apart from those recruited from the same institution, people came into the network as strangers, but with a new basis for interaction, which was the network itself. As Roy Gravel explained,

> I didn't know who Michael Hayden was, and I didn't know many of the scientists who subsequently became involved. It wasn't so much that people stayed on one side of the continent or another. It was just harder to find people throughout Canada. So this network idea became interesting very quickly, because we met new people doing collaterally related things.

The recruitment process was quite divisive, however. The rights and wrongs of who was, and was not, invited to join are still being debated. Four levels of investigator were specified in the proposal. Six of the original 'group of seven' were designated *principal investigators* (PIs) – individuals with 'established international reputations' in the field of molecular and/or human genetics. All of the PIs were men, three of the six were based at Sick Kids; two were from McGill, while Hayden was the sole representative from the west. The seven scientists at the next level were designated *research associates.* These four women and three men (one from the founding group) were individuals with 'established reputations' in human genetics, many of whom were shifting their research program to the molecular level. Of the seven, four were based at Sick Kids, one was at McGill, and two represented the prairies. Hayden was still the sole representative of UBC, the headquarters institution.

A third level was called *young investigators*: all men, these three young Canadian scientists – one each from the universities of Ottawa, Montréal, and British Columbia – were said to have demonstrated 'outstanding creativity' in the early stages of their career. The significance of the final level – *core facilities directors* – was immediately understood by Hayden but perhaps not by the others at that time. Directed by four men and one woman, the core facilities quickly became the key to the network's success. In fact, the core facilities came to define what it meant to do 'network science' – they were true 'collaboratories.'[9] Core facilities had both cognitive (human) and material (non-human) elements. They were a combination of the directors' technical expertise and interventions and the material equipment and instrumentation. Because Hayden realized the importance of these advanced technologies, directors of three of the five core facilities specified in the proposal were based at UBC.

In all, the twenty-one scientists listed as network members in the 1988 funding proposal represented eight universities and five university hospitals and/or research institutes: University of British Columbia (including the University Hospital and the Biotechnology Research Centre), University of Calgary, University of Toronto (including the Hospital for Sick Children), McGill University, Université de Montréal (including Hôpital de Ste Justine), University of Ottawa (including Children's Hospital of Eastern Ontario), Queen's University, and University of Manitoba. Table 3.1 summarizes the investigators by level, their institutions and locations, and their research interests. (See also table 4.1 for comparison with Phase III.)

The last few days before the submission deadline for the proposal were especially intense. In the words of Ron Worton, it was 'an enormous effort':

> I flew with my secretary to Vancouver for the last six days or so before the proposal was due, because it was too awkward to try to manage it from two cities. And this was the early days of computers, they were fairly crude at that time. Their memories were small. But Excel had just become available ... So, we went out and bought that program a couple of days before I flew to Vancouver. My secretary was reading the Excel manual on the airplane so that when we got to Vancouver, she could do all the spreadsheet work to put the budgets together.

With everyone working around the clock the proposal was submitted on time, 30 November 1988, under the title *Genetic Basis for Human Disease: Innovations for Health Care.* It was one of some 158 formal proposals submitted in response to the original call. The leadership issue had been decided by then. Hayden would be scientific director and, by virtue of that fact, the University of British Columbia would host the network's administrative offices. Worton and Scriver were listed as co-directors.

After the excitement subsided, everyone went back to their labs while the process worked its way through the bureaucracy. Given the intense activity of 1988, the hiatus was something of an anticlimax. It took almost a year before the successful networks were announced. Then, on 26 October 1989, the fifteen networks were notified of their awards. The genetics network would receive $17.5 million over four years.[10] Asked why he thought the CGDN proposal succeeded, one of the founders responded,

TABLE 3.1
CGDN investigators listed in 1988 proposal for Phase I of NCE program

Name	Institution	City	Research interests
PRINCIPAL INVESTIGATORS			
Gravel[1]	HSC/UT	Toronto	inherited biochemical disorders including Tay-Sachs
Hayden	UBC	Vancouver	late onset genetic disorders including Huntington's
Scriver*	McGill	Montreal	physiological genetics and human genetic variation
Skamene	McGill	Montreal	genetic susceptibility to disease
Tsui*	HSC/UT	Toronto	cystic fibrosis and gene regulation
Worton*[2]	HSC/UT	Toronto	Duchenne muscular dystrophy and genome structure/function
RESEARCH ASSOCIATES			
Cox[3]	HSC/UT	Toronto	antitrypsin deficiency and human genetic variations
Field[4]	UC	Calgary	genetics of multifactorial disease including diabetes
Gallie	HSC/UT	Toronto	retinoblastoma and other genetic malignancies
Greenberg[5]	UManitoba	Winnipeg	hypophosphatasia
McInnes	HSC/UT	Toronto	genetic diseases of the retina and inherited biochemical disorders
Morgan*	McGill	Montreal	complex phenotypes and population genetics
Robinson	HSC/UT	Toronto	lacticacidemias
YOUNG INVESTIGATORS			
Goodfellow[6]	UBC	Vancouver	multiple endocrine neoplasia
Korneluk	CHEO/UO	Ottawa	myotonic dystrophy
Mitchell	HSJ/UM	Montreal	inherited biochemical disorders
CORE FACILITIES DIRECTORS			
Aebersold[7]	UBC	Vancouver	protein analysis and sequencing
Duncan[5]	Queen's	Kingston	in situ gene mapping
Jirik[8]	UBC	Vancouver	transgenic mice and gene targeting
Lea[6]	UT	Toronto	hybridoma technology
Lee	UBC	Vancouver	electron microscopy

* Also core facilities directors
1 Relocated to University of Calgary, 1999
2 Relocated to University of Ottawa, Children's Hospital of Eastern Ontario, 1996
3 Relocated to University of Alberta, 1996
4 Relocated to University of British Columbia, 2001
5 Until 1996
6 Resigned 1992
7 Resigned 1994
8 Relocated to University of Calgary, 2000

> The NCE review committees looked at our science, first. That's your ticket to get in. Once you've accomplished that, you also have to demonstrate that you have a different outlook within the network than in the basic science system. So, the balance I thought was good. We had the breadth of everything.

For the new networks, the nine months following the announcement of Phase I awards – the gestation period from November 1989 through July 1990 – were chaotic, as federal bureaucrats struggled to put administrative structures in place. The first tranche of funding was not advanced until August 1990, more than three years after the program was first announced as part of the April 1987 InnovAction strategy, and thirty months after the funding commitment was made in January 1988. The delays indicate the novelty of the program. Federal systems to implement and manage it had to be developed *de novo*. In the selection process, most of the attention had been paid to scientific excellence. In the implementation process, consideration had to be given to the other criteria: linkages and networking; relevance to future industrial competitiveness; and administrative and management capability. These non-scientific elements constituted a large part of the program's novelty. Taken together, they meant NCEs would function as 'research economies' with proper management and governance.

These elements would be covered by a 'memorandum of understanding,' as it was then called, an internal agreement governing each network's formal 'powers of association' – its management and governance structure and its public- and private-sector partnerships. (Note, however, that legal powers resided with the host universities.) CGDN's first internal agreement was signed on 4 July 1990. Two industry partners were signatories – MDS Health Group Ltd (later MDs Health Ventures Inc.) and Merck Frosst Canada Inc. – as well as the thirteen institutional partners referred to earlier. (However, where researchers worked in university hospitals, both the university and the hospital were named as network partners, making the network appear more extensive than it was.)

Once the formal agreement was in place, funding was released and the network could seek staff to fulfil the non-scientific criteria. The dynamics of network formation came into play here, too. The network's administrative manager was recruited from industry partner Merck Frosst's research planning division in Montreal. She set up the initial systems. Then Dr David Shindler – a leading science policy adviser – was recruited from Canada's Science Secretariat in London as the network's managing director, following an approach by a network researcher. With the two key employees in place, the network's administrative centre was opened at UBC in September 1990.

Managing the Network

A former program officer, now a policy adviser, described the history of the NCE program as 'the evolution from free research to managed research to industrial participation.' The NCE directorate believed that management expertise and governance could make or break the networks and was as important as the excellence of the science. Management would be one of the key features that distinguished networks from academic science-as-usual. As stated earlier, the NCE program was conceived as large-scale managed research. In the post-genomics era, managed research is routine. At the time, though, it was 'a major novelty and a shock to many' said a key federal source; 'perhaps it was the first culture shock.' Not every one felt the same, however; an Industry Canada informant views NCEs as simply 'slightly more managed or administered' than is usually the case in academic science; managers were there to look after the paperwork, knock on doors looking for partners, and otherwise free researchers from tasks that diminished their productivity. The two interpretations – 'culture shock' and 'normal practice' – reflect the cultural differences between the program's governing agencies: the research councils on the one hand, and Industry Canada on the other.

Tensions within the System

Stipulations were put in place that all networks would have a board of directors, a scientific committee to organize the research program, and a management team. Network boards and committees were to be structured to bring the expertise of industrial partners to bear on research management. This industrial representation took time to achieve, however. In Phase I, CGDN's board was heavily weighted to academics, and UBC's dean of medicine was chair.

The federal decision to restructure the selection criteria for the Phase II competition put the scientific and non-scientific mandates on a par. This decision reflected Industry Canada's concern that, in Phase I, too much emphasis had been placed on research *excellence* and not enough on industrial *relevance*. In other words, unless formal management structures were given equal status, it was far too easy for a network to allow researchers to do 'science-as-usual,' that is, to follow serendipitous directions and do more or less what they wished to do. The pressure for increased management was also a function of the increasing size of network research programs.

A program officer suggested that the managed research model brought an overall, strategic vision of a type unusual in academe. As CGDN's senior executives interpreted the management mandate, 'some level of cohesion, some level of network identity, some level of management, some level of cooperation' was required. In a distributed network, where people do not necessarily see each other, 'there has to be some [management] glue at the core; if there's no glue there it ain't going to work.'

But CGDN scientists were not used to being monitored by managers. At least in Phase I, network funding looked to them like business as usual: just another federal grant. It was the task of management to persuade them otherwise – that not only the standard of excellence but *all* of the program's mandate requirements had to be met. Managers made the baselines clear.

> If you fell down on any one of them, you were finished ... We had to be pretty tough, and it was hard. It was painful. We had to kick people out of the network when the work wasn't up to scratch. When they didn't maintain their science, or they weren't doing it the way we saw it had to be done.

This level of control over researchers was possible only because the most senior executives held PhDs. Both the original managing director and his successor belonged to the scientific culture and enjoyed peer status with network researchers. Their scientific credentials helped to establish their credibility when enforcing accountability. As members of the culture, they understood the competitive nature of scientific careers. While they reinforced high standards and the orientation to excellence, they also sought to encourage researchers to maintain their science and be acknowledged for it. 'It wasn't just about grants,' a senior executive noted, 'but to be recognized by their peers for the good work that they were doing.'

The maintenance of standards paid dividends. By adjusting to the program's changing demands, CGDN won a total of fourteen years' funding in all, the maximum allowable. The network was successful in each competition, being renewed for Phase II in the mid-1990s, and again for Phase III in the late 1990s. After Phase I, much of the hierarchical partitioning of researchers disappeared. In subsequent competitions, all the original associates were reclassified as principal investigators (PIs). Core facilities directors were also listed as principals, reflecting the reality that most ran research programs as well as providing a service to other members. The category of 'junior researcher' disappeared. (Subsequently, promising young researchers were appointed as 'network scholars' on a fixed term.) Network documents show thirty-three PIs at the start of Phase II, representing nine universities and four related hospitals/institutes. After the Phase III expansion, the network agreement details fifty PIs at twelve universities and eight related hospitals/institutes.[11]

The hospitals and universities were rarely enthusiastic signatories to the network agreements. To them, a network was a problematic organizational entity. Given that Ottawa's original intent was to bypass university autonomy, it was little wonder that conflicts occurred between these reluctant 'hosts' and their unwanted 'guests,' as the networks established their institutional identity. As Michael Hayden described the relationship, 'Universities didn't trust the networks. They saw us as a power grab. They saw too much power going to the networks away from the universities. And they didn't trust and didn't understand the process.'

Institutional Friction

Universities and hospitals that house network offices and researchers are called 'host institutions,' but their hospitality is largely involuntary. The legal status of networks is an important factor in understanding the host-network relationship. Under corporate law, collectives (e.g., societies or associations) hold certain 'powers of association' not available to members as individuals. Those powers are exercised through the association's officers, professional staff, and governance mechanisms. Legal powers of association, and legal personhood, require incorporation, and CGDN did not incorporate until 1998. Until then, in legal terms, it did not exist.[12] As a CGDN manager said, 'These are very fragile organizations; they're built on practically nothing. There is very little holding them together except money.' Until 1998, then, CGDN was an 'ephemeral organization'[13] existing only in the interstices of university accounting systems. Its status in relation to the university was highly ambiguous.

Commenting on the network's location on the periphery, Michael Hayden said, 'We were federal but we weren't in the mainstream. It was strange.' But from the margins, as 'federal agents,' NCEs were able to mobilize significant *informal* powers of association. In the absence of formal identity, they bound CGDN together with a *willed* identity. In the words of a senior executive,

> When you are a network, when you're not incorporated, when you're undefined, when you're an instrument of the university – the universities consider you their instrument even though you are not – and when you're trying to do something *in between* everybody else, it's very difficult to establish an identity. And we worked hard to create an identity.

As the networks developed distinct identities, two clear sources of friction with host institutions emerged. The first source of friction was the financial costs of hosting networks. Unlike the National Institutes of Health in the United States, Canada at that time had never funded infrastructure costs for medical research and only rarely allowed researchers to charge their salaries to research grants. Whenever a new program was established, universities had to cover the additional costs. By any standard, NCE overheads were large and expensive for university budgets to absorb. In effect, these institutions supplied the incubation facilities in which networks could flourish but received no compensation from the program, or recognition for their contribution. As well, it was a case of 'taxation without representation' since universities had no power to regulate the activities of networks, which were accountable only to Ottawa.[14]

Overall public investment in the program from fiscal 1991 to fiscal 2001 approached $759 million (see table 3.2). But this figure does not include university infrastructure or the salaries and benefits of university researchers. The NCE directorate conservatively estimated the latter at approximately $100 million a year in 1996.[15] Using the growth of the program since 1996 as a base, we can calculate that the annual salary figure had likely doubled to approximately $200 million a year in 2001. According to one federal informant, by absorbing these costs universities have contributed at least as much as the program itself over the years. Acknowledging the historical underreporting of public support for the program, the director of the NCE program estimated that

TABLE 3.2
Total cash contributions to NCEs, 1991–2001 (excluding in-kind gifts and overhead support)

Agency	C$ million	%
NCE grants	586.9	69.0
Federal agencies	39.6	4.7
Administration/sundry	14.2	1.7
SUBTOTAL – FEDERAL	640.7	75.3
Provincial agencies	52.6	6.2
SUBTOTAL – GOVERNMENT	693.3	81.5
Universities (direct only)	10.5	1.2
Other – hospitals and tax-exempt foundations	55.2	6.5
SUBTOTAL – PUBLIC SUPPORTED INSTITUTES	759.0	89.2
Industry contributions	91.5	10.8
TOTAL CASH	850.5	100.0

Source: Compiled from NCE annual reports, 1990–2001

'the additional contributions from both the granting councils[16] ... and the universities tends to almost triple the total amount.'

In contrast, over the same period the private sector is credited with only $91.5 million, or approximately 11 per cent of the program's 'official' $850 million cash budget. Even this figure is overstated as a result of various reporting anomalies regarding cash contributions that will be discussed later. The same anomalies prevent any reliable estimate of 'in-kind' contributions from industry partners. Without full estimates of cost, it is hard to calculate the program's cost-benefit ratio.

Perhaps understandably, universities resented their expensive and uncontrollable guests and did what they could to assert their institutional authority. According to one CGDN informant, the initial reaction was, 'The government has forced these damned networks on us ... why should we even talk to these network guys? What do they bring to the table?' One way for universities to manage the intruders was through bureaucratic controls. As already stated, prior to incorporation CGDN had no legal capacity to hire employees, make contracts, or receive funds. In all such arrangements the host university acted as surrogate, as if the network were a minor child, incapable of forming intent. CGDN's researchers were employed by their individual hospitals and universities; network staff worked for UBC. When CGDN wanted to hire David Shindler as managing director in 1990, UBC refused. As Michael Hayden recalled,

> He was the guy we wanted, [but] the only way to hire him was *not* to hire him but to get him to take a secondment from his current job. We would pay the Ministry of Foreign Affairs and they would pay him. We did that for five years. It took UBC that long to approve the appointment ... [to] become more trusting of the networks.

The universities wanted the networks brought under university control. As one network manager described the situation,

> This was about power and greed. They wanted control of our budget. They wanted the ability to claim that the networks came under the universities, so that anything the networks accomplished could be attributed to the universities. It's more money for them, it's more profile for them. It's a case of *the bigger our basket is, the more of a power base we have.*

CGDN's principals resisted the administrative blocks imposed by the university. While acknowledging that university budgets were inadequate, they saw no reason to accept the blame and pointed to waste and inefficiencies that 'leaner' structures like networks avoid. 'Universities are under-funded, but they are over-headed,' said one founding member. 'There is too much infrastructure. To lay blame onto the networks for some aspect of it is unfortunate and misplaced.' On the contrary, he suggested, universities should recognize the networks as assets.

A second cause of friction between host universities and networks relates to the management of intellectual property (IP) generated by network researchers. Both networks and universities are involved in what Robert P. Merges has described as a process of 'creeping propertization,' as discoveries that would otherwise have remained in the public domain, are 'captured' (privatized) as intellectual property, then exploited for profit.[17] In this drive to propertize the products of science, NCEs and their host universities compete for profits. Each seeks to depict itself as the most legitimate agent and skilled representative in the drive to turn science towards the market.

Beyond the drive for profit lie several distinct irritants. First, the internal agreements that are supposed to govern intellectual property issues are complex documents universally described as unmanageable and 'ugly.'[18] The agreements are supposed to clarify relationships and IP ownership issues – but they do not. This means that each commercialization deal must be treated on a 'one-off,' case-by-case basis.

Second, over time networks have become more aggressive about intellectual property. Initially, during Phase I, the NCEs provoked fairly limited interest because program demands in this regard were modest. Phase II brought increased expectations on the part of the program and a matching response from the networks. Since Phase III, the networks have been looking to IP commercialization to carry them beyond the sunset of NCE funding. As one university technology manager commented, 'The networks are really fighting for our intellectual property ... the reality is that if they're going to be self-sustaining, they have to insert themselves into the process.' Another said, '[These] people are trying to protect their future at our expense.'

Finally, there is a sectoral disparity among the networks in their ability to deliver commercialization services, and in their approach to technology transfer. According to university technology managers, the information technologies and electronics networks tend to be 'fairly hands off and laissez-faire,' whereas the life sciences networks like CGDN tend to be proprietary and centralized. Because life sciences, networks control their boundaries and members, they have been able to make themselves 'obligatory passage points' for IP protection in a way that university commercialization offices have not.[19] In networks, the processes of *interessement* and *translation* ensure that discoveries with commercial potential are disclosed to the network first.

Industry liaison offices (ILOs) in universities argue that NCEs duplicate existing technology transfer infrastructure and add little value in the process. In turn, the networks point out that, historically, universities had no incentive to pursue commercialization nor any particular interest in doing so. One of the driving forces behind the establishment of the NCE program, they say, was to 'leach out' technologies otherwise languishing in universities.

University technology managers argue that they carry most of the workload for the development of NCE technologies while receiving little credit. One manager stated, 'On any technologies that I've been dealing with NCEs, I would say I've done eighty per cent of the stick handling.' But to a CGDN board member (private sector), university ILOs 'appeared to be uniformly inept or non-existent or both. The networks were much more competent.'

In comparison with 'Johnny-come-lately' narrowly focused networks, ILOs depict themselves as deeply experienced and possessing a 'whole university' vision. In contrast, networks hold themselves out as fast-moving sectoral specialists, moving strategically to secure IP. They depict ILOs as lumbering, bureaucracy-bound generalists with no industry experience, trying to handle everything from astrophysics to zoology. According to a former CGDN commercial director, ILO staff just don't develop a good understanding of how industry thinks, so they don't really understand how to find market prospects: 'They mean well, and they try hard and they work hard. They often are extremely overworked for what they get paid. But, you know, we were focused on our own field. And that meant we could specialize.'

Heroic tales are told about the relative competence of the network and the ineptitude of the ILOs. These myths have entered the collective belief system and seem to be part of the enculturation process. A classic example is CGDN's Alzheimer's genes legend, a story that is repeated, in various forms, by board members, researchers, and professional staff. The discovery of two genes for early-onset Alzheimer's disease was a big find. The university was not willing to move fast enough to protect the IP rights, so the network took the lead, realizing, said a network manager, that 'if we didn't patent it – yesterday! – we'd lose it.' The legend describes how the heroic managing director got the genes patented within forty-eight hours, thereby protecting the discovery's commercial potential for Canada. As recounted by Ron Worton, the network's associate scientific director, the authorized version of the tale is as follows:

> This was well into the NCE process – by now we're talking about Phase II and we're into about the winter of 1995. The researcher called me one day and said, 'You know, we've got the Alzheimer's gene finally. I've gone to the university and they don't think that it's worth patenting. They don't think that it's worth anything. They don't want to follow up on it. What should I do? Do you think the network would be interested in helping me to patent it?'
>
> So I called the network's managing director in Vancouver five minutes later and said, 'You've got to call this guy and talk to him about the patenting. The university will be convinced that they need to be involved in the end, but would you take a lead role here and at least make sure that he doesn't go out and publish the stuff before it gets patented?'
>
> And the managing director said he would do something. That was like five in the afternoon. Ten o'clock the next morning, he phones me back. He is in Toronto, walking down University Avenue, talking to me on his cell phone. He'd flown in on the red-eye overnight, set up a meeting with the researcher for that morning, and by mid-afternoon, they were well on

> their way to developing the patent position and talking about the whole strategy for exploiting this intellectual property. And of course, as soon as he got involved, the university realized that there really was something there that they should be involved in. And in the end, it worked out well for everybody. But, I think that was the first time I had seen the network really play a catalytic role in making something happen.

Ultimately, this initiative resulted in what was, at the time, the largest IP deal in Canadian university history, between Schering Canada Inc. and the University of Toronto in 1997. Schering's initial $9 million funded a three-year research program in the development of drugs and technologies to treat and prevent Alzheimer's. Over the long term, the agreement carried a potential value of $34.5 million, not including royalties.

Despite the sniping about the relative levels of commercial competence, network researchers work not in 'networks' but in universities and hospitals, which pay their salaries, provide their lab space, and pay their overhead and operating costs. Resulting technologies are owned by the institutions. Their ownership of IP is 'cast in stone' and they are not about to cede their interests to the networks.[20] Thus, networks and universities *must* work together, or nobody benefits. In game theory terms, it is a classic prisoner's dilemma. Over time, both have made concessions, and a truce of sorts has been worked out.

Spatial-Structural Dynamics

Reports of CGDN's configuration are rife with power relations and exclusionary criteria. These issues are best understood as relating to the network's spatial-structural dynamics: the larger 'why' questions of regional distribution, elitism and equity, social reflexivity, and fiscal accountability.

Regional Distribution

As befits a federal program, success in fostering a wide *national* distribution of networks and resources is a policy concern. However, the experience of CGDN shows that this goal may not be realistic. When the program was being planned, the 'network' component appealed to politicians because it offset the elitism implied by 'excellence.' To a Canadian politician, elitism means geographical concentration.

TABLE 3.3
CGDN funding allocations by institution, 1991–2000

Institution	$ million	%
University of British Columbia, Vancouver	7,076,592	21.1
Hospital for Sick Children (University of Toronto), Toronto	9,486,464	28.3
McGill University, Montreal	7,343,483	21.9
All others	9,590,292	28.7
	33,496,831	100.0

Source: Compiled from CGDN financial records

According to a federal informant, the program was sold to Cabinet 'as an economic development package – a regional economic package. But Cabinet was sold "a bill of goods."'

Despite rhetorical claims of national scope, and a significant expansion in Phase III, CGDN's main clusters remained at the three original institutions: Vancouver's University of British Columbia, the University of Toronto's Hospital for Sick Children, and McGill University in Montreal. An examination of research and core facility funding allocated to network PIs shows that these three institutions commanded more than 70 per cent of the network's $33.5 million research budget in the period 1991–2000 (see table 3.3). Looking at the provincial distribution of network funding in the same period, PIs in British Columbia received 22 per cent, those in Ontario got 43 per cent, while researchers in Quebec received 27 per cent. The remaining 8 per cent was allocated across all other provinces.

These figures indicate that the network is tri-nodal rather than widely distributed. Thus, the 'network' metaphor is misleading; 'spokes and hubs' may be a more accurate descriptor (see figure 3.1). Well-established researchers and locations are favoured.

Actor-network theory relates the density of linkages in particular areas to the activities of spokespersons and their success at *interessement* and enrolment. This is certainly the case with CGDN. The tacit or embodied aspect – the 'spokesperson factor' – can be clearly seen when established researchers relocate to another university. New clusters begin to form around them, confirming the importance of face-to-face interactions. When Diane Cox relocated from Sick Kids to the University of Alberta in 1996, the university had no network members. By 2001, three PIs as well as several associates were based in Edmonton. In the same year, Ron Worton moved from Sick Kids to Ottawa, where

Figure 3.1: Tri-nodal distribution of CGDN funding, 1991–2000

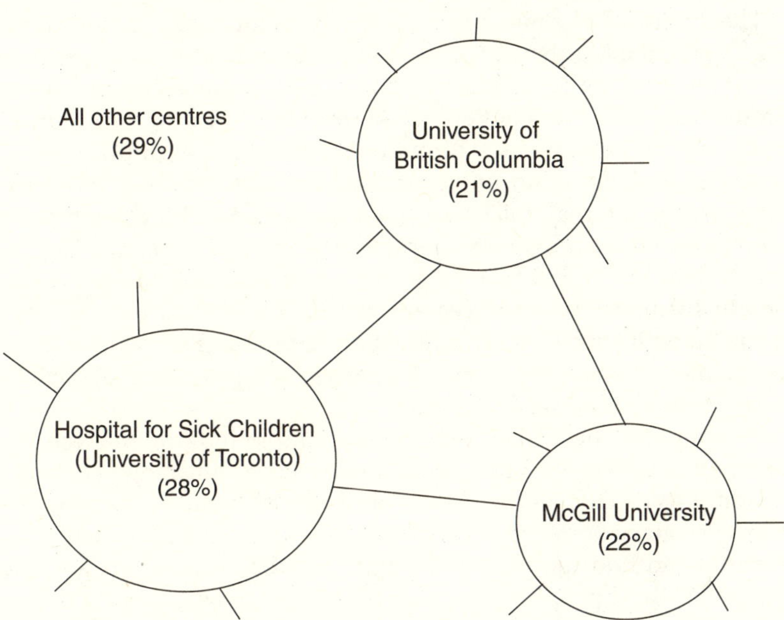

Bob Korneluk was the sole representative of the network. The University of Ottawa subsequently became a significant node, and StemNet, the stem-cell research NCE founded by Worton in 2001, is headquartered there. Finally, Leigh Field was for many years the solitary network researcher at the University of Calgary until Roy Gravel moved there from McGill in 1999, followed by Frank Jirik in 2000. (Field relocated to UBC in 2001.)

Other spatial and structural factors must be accounted for as well: for example, proximity effects and institutional context. The significance of space and proximity ('territorialization') is a commonplace in economic geography, which accounts for the ways resources and tacit forms of knowledge are generated and shared. Regional clustering of researchers, institutions, and firms is a well-recognized phenomenon in the literature on industrial districts and regional systems of innovation.[21] As Jonathan Murdoch notes, 'Networks are differentially embedded in particular places and ... different forms of organization evolve in different sociocultural contexts.'[22] Something similar occurs in CGDN: the combination of inertia and proximity means that it is easier to build linkages with researchers in the same or nearby institutions than with those at a distance.

The institutional context is another key factor in facilitating clustering. Again, the Matthew effect is at work. One institution begets more. These institutions layer together to create a regional system for the production and exploitation of knowledge, creating 'institutional thickness.'[23] The network's Toronto node is a good example: there, six hospitals and the main university campus are within steps of each other. But an internal 'thickness' is also important: laboratory practices are shaped by the university's formal structure and context.[24] This context defines the 'rules of the game': for example, determining the ways in which university resources are allocated and who can command them. Some institutions concentrate more power than others and can assign more resources to particular enterprises, providing a hospitable environment for network activities.

In summary, the CGDN example appears to indicate that the NCE program supports the institutional status quo by directing resources to existing research 'centres'; the 'peripheries' remain marginalized. However, since knowledge is embodied, it appears that if smaller universities can find the means to attract network researchers and their programs, these people become agents of change who attract others.

Elitism: Norms of Equity and Exclusion

The concept of centres and peripheries is closely linked to that of inclusion and exclusion. Both are cultural oppositions, linked to spatial notions of familiar and strange, presence and absence. Examining the norms guiding enrolment may help to reveal why some 'strangers' were included in the network while others were excluded.

The first International Peer Review Committee was 'unapologetically elitist,' according to a committee member; the term 'excellence' meant that the NCEs 'should pull together world-class teams of scientists: the very best people who, with support, could pull the rest forward.' Roy Gravel recalled that 'excellence' was defined as the top 5 per cent of scientists in a field, worldwide. He considered that an odd statement, 'because science doesn't work that way. That wouldn't be the way you would identify the cream of Canadian science. And that wasn't a Canadian number ... so it had no meaning.' Nevertheless, given the 'excellence' requirement, the biggest challenge in putting the proposal together was choosing the people.

The core-set had to ensure that program requirements (for example, geographic distribution) were satisfied, while covering the domains of science that interested them – human genetics, medical genetics, and key technologies. But the NCE program's preference for Mode 2-type interdisciplinarity was largely ignored. Early in the planning, says a founder, the core-set decided 'that this would be a network of molecular geneticists. And so anybody who was doing cybergenetics, or biochemical genetics or any other type of genetics [was] automatically excluded in order to keep it focused.' This network would operate almost entirely within traditional disciplinary bounds.

Beyond that, a degree of arbitrariness and capriciousness prevailed in the debates about who the core-set did and did not want to work with. Perhaps this was inevitable, given the need to select only a couple of dozen people from across the country. However, in designating a handful of people as superior scientists who were 'excellent' enough to be in the network, they left an impression that those who had been excluded were somehow inferior. The process left a legacy of ill-feeling. Lap Chee Tsui regretted the elitist direction: 'In retrospect, I think we should have included everyone. The whole community is very small, and in the end about seventy-five per cent became a part of the network. So there was a small number of people who did not get in. I just felt it wasn't really necessary to go through the agony when the numbers were so small.' Ron Worton described the process:

> The biggest challenge was not in determining who we should choose, but who we should not choose. We made that determination with difficulty, and somewhat arbitrarily. There were some pretty good scientists in the country that we excluded ... For whatever reason. Maybe we felt their publication rate wasn't high enough, or they weren't well enough known, or we didn't like the way they did their science, so we excluded them. And in the early days I got phone calls from some of my friends who said, 'I'm really angry that you guys did not include me in the network. Why did you not include me?' And when you're asked a question like that, it's almost impossible to answer. It's about standards and focus really.

The network was a kind of elite club where membership was increased by invitation only. The inner circle – the priorities and planning committee – would 'throw names on the table and discuss them,' according to a committee member. Often, names were put forward by other researchers, but even with those bona fides not many were chosen. Few outside the inner circle understood the selection process. According to Worton, the committee attempted to identify people whose research looked 'interesting' and complemented the existing research program.

One member, a junior researcher back in 1988, thought the decision to include him in the network was circumstantial. He had trained at Sick Kids under Roy Gravel and was located in Ottawa, which gave the network an opportunity to add a node beyond the Vancouver-Montreal-Toronto triangle. He said, 'They tried to cover all possible aspects. Scientists in different parts of the country. Scientists that were young and scientists with a lot of experience. So when they went down the list, I guess I ended up included.' He recalled that Mike Hayden used to joke that they needed at least one person in Ottawa to deliver the funding proposals.

For a similar reason – to get wider geographical representation – Hayden contacted a researcher at the University of Manitoba who would represent genetics researchers on the prairies. Another, at the University of Calgary, was self-selected: 'I heard they were doing this and I wrote them a letter, I guess it was to Mike Hayden, and said I'd like to be part of it. And he said, "Well, send me your CV," and I did, and they invited me in.' One person from the group at Toronto's Sick Kids remembered that 'it was initially extremely exclusive. And then it widened out a little bit to include those people who had a particularly high ranking in MRC, and I was one of those.'

In a 1997 opinion piece in the journal *Nature*, a molecular biologist and a zoologist argued that the life sciences are in danger of losing their originality.[25] The authors perceived a homogenization of opinion, with fewer independent schools of scientists finding novel approaches to problem solving. They maintained that scientists are 'playing safe' by following established lines of enquiry, rather than taking intellectual risks. The authors believed that this situation was perpetuated, in part, by the dominance of 'star' scientists at conferences and in the literature, and in the inherent conservatism of the peer-review process. Extending that analysis to the NCE program, these networks, by limiting selection to elite scientists, tend to limit the variety that feeds more risky, innovation-led research.

Another anomaly concerns gender. Women PIs claimed that the role of female scientists in the network had always been equivocal.[26] Only five of the original twenty-one members were women, and all the original PIs were men. Of the total research and core facility funding allocated in the period 1991–2000, women researchers received 11 per cent rather than a proportional 24 per cent. The two founder members who were not renewed in 1996 were both women. As one of the five female founders pointed out, 'Some very senior women scientists were not in the network at all. They were not invited.' The proportion of women increased slightly over the years. In the Phase III proposal, submitted in 1997, eight were listed as members. In 2001, one of three new PIs was a woman, as were three of five new junior researchers known as 'network scholars.'

All five women founders referred to the network's underrepresentation of females. Most saw the problem as systemic rather than specific to the network but expressed some exasperation about the general lack of concern shown by the network's male core-set. In addition to the frustration with gross numbers was the fact that women were not represented in power positions:

> It was very strongly male dominated. And we [women] have had little involvement in [running] the organization. I'm not even sure that [the men] notice, particularly. The women used to joke about it. But there's a problem that way in our field, in Canada, in general. There's a core of people who are very supportive of each other, in and out of the network. And it's very difficult because you're not a 'buddy' of the guys. I'm not suggesting it is a major complaint or anything, but it's simply a fact. I think it's better now for the younger investigators in the network ... but the senior women are scarce.

The network made no serious effort to attract females, said another, 'even though there is a lower percentage of women in the network than is generally the case in human genetics in North America.' A 2001 report by the U.S. National Science Foundation tends to support this assertion: unlike the situation in the physical sciences, about half the doctorates in biology are awarded to women. Even in the 1980s, women earned one in three biology doctorates.[27] Thus, it is curious that all of the individuals who were dropped from the network were women. One explained that it is simply much harder for a woman to succeed in medicine and science: 'The nature of the [science] system is that it's run by men. If women ran the system it would be very different. So there is no question there is a sexist component to it. It is just because men make the rules.'

The women found the elitist 'invitation only' approach particularly troubling and complained about the lack of transparency in the selection process. They could find no logical explanation for who was 'in' and who was 'out,' for either gender. Names had been proposed, but to little effect. Said one, 'I don't exactly know what happened to those suggestions, but apparently they were looked at by the [leaders] who decided not to invite them.' People who were excluded found a question mark hovering over their career, especially as the network grew in academic prestige. A woman manager noted that, 'People began to wonder ... "Why didn't they invite me," you know? It's the coalition of top geneticists in Canada: "Well, why haven't they invited me?"' When the network was just starting, being passed over carried little significance. But the larger the network became, the worse it was to be left out. Certainly, several well-known Canadian geneticists were excluded. One said, 'It gave the impression then, and probably still does today, of being a kind of an elitist club, and one in which I didn't belong.'

With one notable exception, visible minorities were also conspicuous by their absence. The network's board, its scientific and professional leadership, and its principal investigators were uniformly Caucasian. Whether or not this uniformity reflects the field of medical genetics as a whole, the homogeneity of race and gender perhaps indicates a profound social, if not scientific, conservatism at the heart of this network. This conservatism was also reflected in CGDN's absence of social reflexivity and public accountability.

Accountability as Social Reflexivity

One of the defining elements of new network forms of organization is said to be their social reflexivity. Rather than being accountable to the community of science, these networks are intended to be accountable to the community at large. It is a pluralist framework, in which the pushes and pulls of the agendas of a variety of social actors condition the network's decisions and policies. Thus, Michael Gibbons and colleagues argue that public interest groups, lawyers, and social scientists have a voice in the governance of Mode 2 research networks and, more controversially, in the composition of research teams.[28] This broad representation is deemed essential because of the risks and issues inherent in contemporary science and technology. Similarly, Michel Callon has noted the emergence of 'knowledge co-production' models, in which patient groups establish themselves as 'partner associations' with research groups and establish parity between lay and expert knowledge of the disease process.[29]

Bruno Latour also emphasizes the social accountability and reflexivity of network formations. He argues that in a culture of 'open science,' where autonomy is sacrosanct, there is no direct connection between scientific results and the larger societal context. But in the network type of scientific culture, there is a *new deal* with society – a type of collective experiment in which science and society are mutually entangled for mutual benefit. He concludes that 'scientists now have the choice of maintaining a 19th-century ideal of science or elaborating – with all of us – an ideal of research better adjusted to the collective experiment on which we are all embarked.'[30]

Recently, Helga Nowotny, Peter Scott, and Michael Gibbons extended the reflexive elements of their original Mode 2 formulation even farther, arguing that scientific knowledge must be 'socially robust' as well as conventionally 'reliable.' Whereas reliable knowledge has traditionally been produced in cohesive and restricted scientific communities (Mode 1), social robustness depends on 'sprawling socio-scientific constituencies with open frontiers' (Mode 2). Socially robust knowledge is superior to reliable knowledge, they argue, first, because it has been tested and retested in contexts of application, and second, because it is the 'underdetermined' outcome of 'intensive (and continuous) interaction between results and their interpretation, people and environments, applications and implications.' The more open and 'comprehensive' the knowledge community, the more socially robust the knowledge produced.[31] Furthermore,

> public contestation, controversy and conflict ... are not to be shunned on grounds of principle. Rather, they are a sign of a healthy body politic and part of the process of democratization ... Space has to be made for what people want, what their needs are, and ... even contradictory responses and claims.[32]

To the extent that the NCE program was apparently seeking to create the type of networks envisaged by Gibbons, Callon, and Latour, presumably with a broad understanding of public accountability, Michael Hayden's notion of civic science seems impoverished. The construction of the scientific citizen is a far more complex process than Hayden suggests. For Hayden, the subtext seems to be that the public (non-scientists) are useful when mobilized *en masse* but must otherwise be kept at arm's length, lest their ignorance and/or interests impede the research enterprise. This is a classic example of science/non-science demarcation; what Tom Gieryn calls 'boundary work.'[33]

One of the network's board members, for example, commented that 'the public is generally quite ignorant on the subject of genetics. I don't say that with any negative sort of connotations. It is just a fact of the matter. Why would they not be ignorant? It is a very complex science.' Because of their ignorance, it was assumed that the public had nothing to contribute to the network, despite the complex social and ethical issues that accompany research in medical genetics.[34]

This paternalistic assumption that citizens lack the epistemic capacity to judge complex scientific issues has been described as the 'deficit model' of science-society interaction. The deficit model assumes that if useful debate is to occur, the public must first become scientifically literate. In effect, to be taken seriously by scientists, citizens must become quasi-scientists themselves; a common scientific vocabulary is seen as a prerequisite for meaningful public engagement. Yet a burgeoning body of work on public consultation, citizen juries, and consensus models clearly demonstrates the ability of non-specialists to master technical content. The same studies point to the urgency of democratic deliberation if the policy options created by genetics and genomics are to be fully considered.[35]

Categories of 'lay' and 'expert' are socially constructed and thus problematic. At the same time, expert discourses exert normative influence over the public domain, reshaping debate on their own terms. Brian Wynne calls it 'a profoundly unaccountable and unreflexive process.'[36] The demarcation of lay and expert knowledge and interests can be clearly discerned in a recent U.K. study of medical geneticists. These experts depicted science as the 'gold standard' that demarcates *fact* from mere *opinion.* Opinion was the province of 'the lay public' – an undifferentiated mass hampered by lack of logic and swayed by political rhetoric. Fact, in contrast, was the product of solid, value-free scientific research.[37] The following remarks made by a senior network manager (a science PhD) illustrate the prevalence of these lay/expert demarcations. The passionate response was elicited by a question about the potential for appointing a lay member to the board.

> I mean what would a lay [board] member do? They would just ask us what *we* were doing. Well we can't explain that. We don't have time. So

we try to pick intelligent members who at least understand the field a little bit.

Public interest science is just politics. I want to tell you that right now! That's politics and I don't want politics in my network. If somebody has an agenda about organic foods or genetic engineering, I'm not interested. What I am interested in is: are we curing disease? Are we solving a social problem? We're just as capable of looking at the risks and balances as anybody else. But in the end, would you rather have a cure for Alzheimer's or not? Which is better? And people agree that, in the end, finding the cure for Alzheimer's is certainly a greater social good than being in favour [of] or against clinical trials, or animal rights, or whatever.

The fact is that it would have been very disruptive to have grandstanding on the network board. Of any kind. The interests of the organization have to be paramount, not the individual agendas of board members. And if you have a board that has a bunch of people with individual agendas on it – public agendas, private agendas, political agendas – then you are going to have a dysfunctional board and a dysfunctional organization. You are going to lose that cohesiveness that is so important. You're not going to be able to function. Because they are going to block you and then you're not going to be able to carry out your program.

So we had federal program officers on the board; we had foundations, we had industry, we had universities, we had *intelligent* people – medical people, physicians – that were thinking about all of these things.

Apart from the evident paternalism, this network appears compelled to equate the public's legitimate interest in the conduct of the biosciences with anti-science or fringe activities. The reaction is exaggerated: if the public is given a voice, rationality will be lost; when scientific problems arise, we must 'trust the experts' to solve them. This approach to problem solving has been called 'deterministic uncertainty'; it assumes that problems caused by science are reducible only by the application of *more* science. The public is excluded; yet studies show that much mistrust of scientific expertise is rooted in just this type of exclusionary discourse.[38]

In a high-trust, high-risk area such as medical genetics, the absence of external voices within the network means the absence of fundamental questioning as to what might be an appropriate place for genetic approaches to illness. As one prominent critic points out, 'It's a major social hazard that nobody is looking at those ethical, legal, and social questions within CGDN. Because there is an implicit assumption that all this will be good for us. And we need to ask: Will it be?' Langdon Winner, speaking of the politics of technological change, has raised these questions in a wider context. The 'problem of elitism,' according to Winner, is a question of the way powerful actors and groups skew the agenda in ways that favour some social interests while excluding others. The powerful define the rules of the game and the allocation of resources. Winner urges those who study the social aspects of science and technology to ask,

> What about groups that have no voice but that, nevertheless, will be affected by the results of technological change? What of groups that have been suppressed or deliberately excluded? How does one account for potentially important choices that never surface as matters for debate and choice?[39]

Thus, taking CGDN as an example, claims that networks are more publicly accountable appear insupportable. Although the idea of a 'new deal' between science and society is appealing, it is not apparent that the new deal succeeds in its aims. Far from expanding the public sphere, network arrangements may be contributing to the erosion of that sphere.

Accountability as 'Value for Money'

Deficiencies in public accountability, although an important consideration, are only one issue. Deficiencies in fiscal accountability also need to be examined. The NCE program was conceived under a neoliberal agenda of public-sector reform that was fuelled by a rhetoric of fiscal accountability. Results- or performance-based approaches tie funded science to key economic and social outcomes. It seems both responsible and logical to account to the public for the use of public funds. But accountability goes beyond use to value. Asking if money is 'well spent' involves asking if it is effectively spent and if it could be more effectively spent elsewhere. To put it specifically, are programs delivering value for money? Can they *demonstrate* cost-effectiveness?[40]

The problem of ensuring that public programs remain accountable and return value for money can be understood as a 'principal/agent' problem of delegation and information asymmetry.[41] The state (principal) delegates the provision of research that will fuel innovation to university scientists (agents), who are induced by incentives (research funding) to comply with the regime of strategic science. But, especially in technical areas, agents always know more about delegated tasks than principals do. How, then, to pierce the veil of technical expertise to ensure that scientists are delivering adequate returns on the state's investments in research?[42]

One solution to this problem is to regulate research performance directly. But that means state control. The neoliberal preference is for refined forms of 'remote control' or steering that induce an internalization of the state's expectations. Through these mechanisms of *governmentality,* 'normalized' subjects come to control *themselves* according to previously established understandings of what constitutes 'the norm.'[43]

Governmentality requires fidelity devices that will measure and induce compliance and provide 'discursive validation' that agents are doing what principals expect them to. Largely, these devices are accounting tools, such as budgets, cost-benefit analyses, ratios and comparisons, statistics, and financial and compliance audits. Accounting tools, however, are far from unproblematic. While appearing impartial, they selectively 'construct the world' from a complex web of social and economic considerations and negotiations. Through its surveillance and control capacities and its ability to determine financial norms, accounting has the power to create a new 'factual' visibility and to discipline performance. Embedded layers of accounting and accountability induce the required compliance.[44]

These types of reporting relationships govern relations between the NCE program (principal) and CGDN's administrators (agent); and between CGDN administrators (principal) and network scientists (agent). The network's head office thus acts as an intermediary that helps assure state-principals that scientist-agents are following the policy agenda.[45] It becomes a 'centre of calculation' for the accumulations of facts to be sent to Ottawa. Following Latour's logic, reported data gather 'positive modalities' and become harder to resist as they move away from their conditions of production (the lab) to the network office (the centre) and then to the program directorate in Ottawa ('centre of centres'). At each stage, data are recombined and reinscribed. The NCE directorate seeks to control the network by specifying what 'makes up' the numbers. But network administrators reinterpret the directions in instructing scientists about what information is to be supplied.

To illustrate, CGDN would report as *network* accomplishments almost everything their (university- and hospital-funded) researchers achieved, from scientific breakthroughs to publications, external grant funding, and the raising of venture capital by researchers in network spin-offs. This over-reporting was so prevalent that many of the 'official' network statistics consulted during this study proved unreliable for the purposes of analysis because they failed to conform to the guidelines established by the NCE directorate. A serious example is that networks were supposed to report, as 'cash contributions' from partners, only funding that flows directly through network accounts. In many cases, CGDN reported funding that went directly to network researchers. The network's legitimate interest in those funds was minimal, but because they flowed to members they were reported to Ottawa as contributions received by the network. Also, researchers were asked to report almost all their research activities as network activities for the annual statistical report. As one complained,

> It seems sort of ridiculous, talking about all of these accomplishments, when in fact you know maybe five per cent of them were funded by the network. And yet they want to hear about all [of them]. So every year I have the same argument, like: 'What do you want me to do? Write what my graduate student did last year? Because that's all that you funded.' And she says, 'Oh, no, put it all in.' And I say, 'Well, why should I?'
>
> And it's gotten, quite frankly, a little bit ridiculous, given the amount of money we get versus the accountability and justification. I mean what do you do? Write your whole program down and attribute it to the network? ... I mean I would say things jokingly like, 'I think we should just spend all the money on ... having great meetings in ski resorts. I'd get more out of it than you pretending to send me money and pay for my student.'

The NCE directorate is not only aware of reporting anomalies but may have contributed to them. For the program as a whole, additional public funding is under-reported, whereas aggregated private-sector contributions – both cash and in kind – are over-reported (see table 3.2). As early as 1938, the U.S. National Resources Committee called such practices 'window dressing.' Today, they are more often called 'spin,' the purpose of which is simply to make results look better than they are, to protect budgetary resources and allocations.[46]

The NCE program's first full-time director, appointed January 2000, said that the problem had been brought to his attention and agreed that 'maybe some better discipline should be followed ... that's something that we will be looking at.' In September 2000, the directorate instituted an audit requirement, with the result that networks now must submit externally audited annual reports. After that directive, CGDN restructured its administration and assigned responsibility for financial and statistical reporting to a new staff member with accounting qualifications.

Summary: The Limitations of Hegemony

CGDN forged an institutional identity and organizational structure under multiple constraints, including demands for both scientific excellence and commercial relevance under managed conditions, resistance from local host institutions, the traditional structure of basic research and the conservatism of researchers, and the sheer novelty of doing something that had never been done before. More than a decade after its launch, CGDN's successes are clear. But, equally clearly, some successes were achieved at the cost of consequences perhaps unintended by program architects.

The concentration of resources in CGDN created a hegemony. The network *defined* the field of medical genetics in Canada. Non-members were 'othered.' Careers were affected. Yet there were no objective criteria for membership. Instead, membership was an 'invitation only' affair, within the arbitrary remit of the same elite inner group of scientists that controlled the network from the start. Power relations were asymmetrical, concentrated in the most powerful actors and in the centre(s) they controlled. It was, quite literally, a self-reproducing 'old boy' network. Relatedly, network resources flowed to the power centres rather than being distributed to scientists across the country. The consequence of exclusion and concentration reduced the diversity of the Canadian 'science system.' As a concomitant, there was no room in the network for lay representations. The public interest was constructed and defined in the abstract, within expert discourses that excluded authentic voices of interested publics.

That being the case, and in the spirit of value-for-money accounting, questions arise about the extent of public investment in the network (as well as in the program more generally) and about the returns on that investment. In the ten years from 1990 to 1999, CGDN's six original PIs received between \$1.4 million and \$1.8 million each in network funding, whereas the fifteen other founders received an average of between \$800,000 and \$1 million. Although these are modest amounts on an average annual basis, it must be remembered that network funding is *incremental* funding. Network researchers also received direct support from non-profit disease foundations, research councils, and industry contracts, while their home institutions underwrote salary and direct costs.[47]

By the time of federal exit, in 2005, CGDN had received more than $60 million in direct NCE program funding. This figure does not include provincial and industry contributions, commercial revenues, or university subsidies to network researchers. It is impossible to tease out of this complex of funding sources the results that are attributable to the network and what would have occurred if there had been no CGDN. The same is true of the program as a whole, in which public investment between 1991 and 2001 amounted to $759 million (see table 3.2). In other words, there is no reliable way to determine whether or not CGDN and the NCE program deliver direct 'value for money.'

4 Culture and Science

The more serious problem with accountability frameworks is that they capture and evaluate only those dimensions that can be quantified, objectified, and made accountable. Non-quantifiable and less tangible practices are *not taken into account*. At the same time, other elements assume new weight *because* they can be quantitatively evaluated, such as the quantity (not quality) of research publications, the numbers of patents held, and the dollar value of research contracts. In short, a focus on readily quantifiable inputs and outputs risks neglecting more complex social variables that resist measurement but are, nevertheless, valid outcomes. The construction of intangibles such as 'network culture' and 'network science' is one of these outcomes. How did the Canadian Genetic Diseases Network (CGDN) forge a scientific culture and community and a scientific legacy?

Small, informal collectives of closely interacting scientists have long been the principal means of scientific advance. Robert Boyle coined the apt term 'invisible college' in the seventeenth century. Then as now, invisible colleges were the informal, interpersonal networks that linked scientists in different locations around shared scientific interests. The natural philosophers of seventeenth-century England, France, and Scotland, for example, enjoyed a prodigious correspondence but sought face-to-face meetings in Oxford and London whenever feasible. Now, however, as Philip Agre comments, the availability of information and computer technologies means that 'invisible colleges are in many ways more visible to the researchers than the physical campuses where they organize their places of work.'[1]

The distributed and informal nature of scientific interaction is also captured in the term 'communities of practice' – self-organizing, self-selecting groups of colleagues whose members are informally bound together by their shared expertise. Communities of practice share a family resemblance with the scientific 'thought-collectives' identified by Ludwik Fleck. These communities, characterized by intellectual interaction and the exchange of ideas, constitute the 'carriers' of a field's knowledge and culture. Similarly, Karin Knorr-Cetina has noted the very different 'epistemic cultures' of molecular biology and high-energy physics.[2]

These concepts help illuminate the development of a distinctive culture and community in CGDN. They also assist in understanding the phenomenon known as 'network science.'

'A Nation of Colleagues'

At the end of its first year of operation, CGDN listed among its achievements the development of 'an ethos and common understanding of what it means to be in a network.'[3] The use of the term 'ethos' indicates an interesting ambivalence. While it draws around the network the cloak of Mertonian ideals relating to the normative structure of science (that is, an ethos comprising communalism, universalism, disinterestedness, and organized scepticism), it also invokes the new ideal of 'network science' with its emergent (counter-) norms, such as patents and industry partnerships. The rhetorical purpose of the claim was to persuade NCE bureaucrats that CGDN took the program's non-scientific requirements seriously. Another claim about network ethos can be found two years later, in the proposal for the second phase of funding: 'We have created a *nationwide department* of human molecular genetics.'[4] The subtext here is a recognition of Ottawa's intent to change the overall research culture in Canada, network by network, by overriding university boundaries and autonomy.

Even assuming that the idea of a network ethos is to be taken seriously, the claims were premature, to say the least. Ethos is a cultural achievement, and the development of culture takes time. As well, an interesting question can be posed about whether culture can be *induced* by imposing a network model or providing funding. But an examination of CGDN's history shows that, very gradually and with a different tenor in each of the three funding phases, a distinctive ethos or esprit de corps did emerge.[5] CGDN's 'induced' epistemic community anchored itself in a discursive space of face-to-face interactions that promoted trust and reduced competition. Scientific Director Michael

Hayden proudly commented, in a research interview, that 'the cooperation and collegiality have just been incredible. It's created a nation of colleagues that is totally unbelievable.'

Inducing Solidarity

Although socialized in 'invisible colleges,' network researchers were confused about and initially resisted the whole concept of 'mandatory networking.' No real agreement suggested what that might be or how it might be accomplished. The network's professional staff had to invent virtual and face-to-face ways of meeting program requirements. They had to grapple with the complexity of somehow linking together a dozen institutions, two-dozen principal investigators, and many postdoctoral fellows and graduate students. Furthermore, the reporting requirements meant that networks couldn't merely state that they were doing networking; they had to prove to the NCE directorate that they were doing it. So ways had to be devised of enticing scientists to comply.

The method the network implemented was to make the funding for principal investigators (PIs) conditional on their participation in network activities. Subsequently, it was hoped, PIs would realize the manifold benefits of voluntary participation. Almost all network researchers interviewed during the study commented on this creative relationship between network funding and network building.

> NETWORK SCIENTIST: Although the other aspects of the network have been much more important, you wouldn't have pulled the people together without the bait of the funding. We would have said, 'I haven't got time to just go and talk with these people.' But you'll go and talk when you know that if you don't, you won't get your funding. And then you find it is really worthwhile having talked to them and it is really fun.

> NETWORK SCIENTIST: The biggest value of the network is not the funds that they give us but the networking opportunities and the collegiality and so on. Although, I have to say that if we didn't have funding for our labs in addition, we'd probably say, 'Oh, I'm so busy, I don't think I'll go to the annual meeting. I don't really need to be there.' Whereas, if we're funded by the network, we have an obligation to be there.

The network funding was not a significant proportion of a network researcher's total budget. Only a small component of the researcher's program would come into the network. Usually, it was the component that would profit best from the collaborative opportunities. Other aspects stayed outside. Even in the early phases, when the network was less extensive, the funding allocated to researchers probably never amounted to much more, on average, than 15 per cent or 20 per cent of their research budget. This would have been enough to support perhaps a senior technician or postdoctoral fellow. According to one researcher, 'Out of perhaps fifteen to twenty projects in my lab, maybe two or three were covered by network funding; the rest were covered by other kinds of funding.'

But a moral obligation was attached to the network funding. Said one PI, 'it got people to buy in to the network concept and become part of it.' The prospect of funding helped them to overcome the resistance to leaving the lab for yet another meeting. And this face-to-face aspect quickly became far more important than the virtual aspects of networking. The latter soon became taken for granted, an enabling technology to further the personal relationships and community of practice that were being forged. As one policy adviser explained,

> The network mechanism ... forced people to get together face to face, because of the funding provided ... Face-to-face meeting is really important, especially early on. You need a lot of personal interaction to make that networking work. And after that you can do it by e-mail and telephone and fax and all the rest of it, but in the beginning you really have to have the face-to-face communication.

The face-to-face community that became the Canadian Genetic Diseases Network began to take shape in 1991, at the first network meeting.

Face-to-Face Community

As the main forum for interactions and exchange, the CGDN's early scientific meetings laid the foundations of network culture and community. Unanimous about the cultural importance of these meetings, scientists considered them one of the main benefits of belonging to the network. The first meeting, held at Whistler, B.C., in May 1991, set the format for those that followed. Because of the NCE requirement to dedicate 10 per cent of the budget to networking, full costs of attendance were covered for PIs and core facilities directors. These people could, in addition, nominate three members of their teams for full subsidy. For example, students and fellows funded by the network or working on network projects could attend free. In rare cases, a technical support member of the group could be included if that member's contribution was deemed to constitute fundamental research. In molecular biology, where rewards usually go to lab leaders, subsidizing conference travel for junior researchers was so unusual as to be unique.

All the participants were expected to present and discuss their results, either through a poster (students, fellows) or an overview lecture (PIs). As a result, delegates to the Whistler meeting faced a hectic three-day schedule of scientific sessions, workshops, and discussion periods. Approximately 100 participants attended from across Canada, including board members, external collaborators, and industry partners, as well as network researchers and special guests. Concurrent workshops debated, among other issues, the topics of 'industrial relationships' and 'the search for the gene.'

This routine may seem much the same as that of any scientific meeting or conference. Scientists get together and give papers as matter of course. But there are significant differences. First, as one of the researchers explained, 'A network provides you with access to a completely different and much broader group of people than you would ordinarily associate with at meetings.' Normally, scientific meetings are segregated by narrow research interest. In contrast, network meetings were broad, covering the field of genetics in Canada.

Second, from the start, the norm was 'full disclosure.' The meetings were intended to encourage in-depth discussion of interesting, early-stage research results, often prior to journal publication. Sensitivity to priority, if nothing else, would have precluded this level of frankness in a 'normal' scientific meeting.

At the same time, however, even in these first meetings, a countervailing force emphasized confidentiality. Unless the participant was a network principal – a researcher or partner (industrial or institutional) listed in the network's internal agreement – he or she was required to sign a confidentiality agreement: intellectual property rights had to be preserved to fulfil the network's commercial mandate. So those 'full and frank' discussions had to take place behind closed doors; participants were advised that discussing results in a closed forum of colleagues did not constitute disclosure for patent purposes. (This is debatable but was not tested.) Even so, researchers were cautioned to apply 'normal discretion in disclosure of scientific data.'[6] In practice, however, it soon became clear that 'normal discretion' was not required.

> NETWORK SCIENTIST: It is totally different than going to a meeting where you have to be careful what you say because someone will rush off and do your experiment and publish it before you get to it.

> NETWORK SCIENTIST: In the network, you're not in competition. And so you can confide and get some valuable feedback from these people, right? It's nice to get up there and maybe brag a bit about the stuff that you've got before it's published. It isn't like you feel, 'I can't say anything because Frank in Vancouver's gonna scoop me.'

> NETWORK MANAGER: It's one of the strengths of this network that we're all in this together. It's difficult out there. The more that you can discuss things, in confidence, the better. You have to be confident that the person you talk to is not going to spill the beans. The trust relationships and the reliance on individual integrity [are] very important.

The third factor that marked these meetings as different from others of a similar type was the social cohesion they engendered. Despite all the scientific gravitas, the social aspects of the meetings remain particularly vivid for most people. Asked what she recalled about the early meetings, one of the founders said, 'We had afternoons off. We went hiking ... We did plays – skits and things. And we had fun.' Few more effective ways can be found to build trust and loyalty – and the foundations of future collaborations – than to play together and build personal relationships.

> NETWORK SCIENTIST: When you know somebody personally, because you've met them at these network meetings, then you are much more liable to approach them, to work with them. It increases the potential for collaboration.

> NETWORK SCIENTIST: I have a strong sense of belonging to the network. What I do is defined within my grant applications. How I feel is defined in my interactions with the network.

A Climate for Collaboration

The network community was about openness and sharing, on the one hand, and building a sense of solidarity and belonging, on the other. Through the annual scientific meetings, everyone in the network knew something about what the rest of the members were doing, and that facilitated a climate for collaboration.

Network scientists became familiar with each other's research from hearing presentations on work in progress. This annual 'overhearing' enabled synergies to occur. As one researcher explained, 'Going to the network meeting, it's a very easy, fast way of getting a survey of who's doing excellent research in Canada in our field. And that ... saves a heck of a lot of time for us all.'

Perhaps listening to somebody talking about a particular gene, a researcher might realize that she had a piece of the same puzzle. Or perhaps she needed to find someone with particular skills to help her with a project. In either case, the two researchers could make contact, confident that their overture would not be rejected. Thus, to borrow a felicitous phrase from one PI, the network acts 'like a blanket purchase order on collaboration':

> NETWORK SCIENTIST: The whole game is sitting open on the table and then you can reach in any direction. Anyone who gets a call from another person within the network has a sense of obligation to talk and participate and collaborate ... It is like asking your brother or sister for something as opposed to someone with whom you don't really have the same relationship. They can't say, 'Sorry, I'm too busy.' Or, 'Sorry, you're competing with me.'

> NETWORK SCIENTIST: I know those people well. I've met them many times at network meetings. I've heard them talk. And if there was anything I needed or wanted, I certainly wouldn't hesitate to pick up the phone and expect that I would get a very positive response.

The fostering of trust and reciprocity on this scale was a unique experience for network scientists, who were more used to a culture of competition than one of cooperation. Reducing competition and enhancing the ability of network scientists to work together provided an advantage to the entire collective.

The absence of competition was, at least in part, a result of the selection process. Researchers were chosen for the complementarity of their programs. No two teams were intended to be working on exactly the same problems. So in the network, as one PI said, 'We're not in competition, because we're doing different things. We're tied together with the common interest, but we are distinct.'

Being collegial also included working for the common good and trusting community decisions. Through the years of meetings and network building, a process of sedimentation took place. CGDN began to settle into the shape it had claimed at the start – a community of colleagues, with a shared ethos and a common understanding of what it means to be in a network. One researcher commented that 'as a group of geneticists we really got to know each other much better than would have happened otherwise.' Another said that the network created value through 'personal contact, personal motivation, driving the science.'

Over time, members began to identify themselves as *network scientists*. Almost by accident, they agreed, government had 'got it right' and produced a capacity to do 'national science.' As Michael Hayden commented, 'It's quite unusual to be led from Ottawa. But this was real leadership.' For Lap Chee Tsui, 'Whether by design or by accident, the federal government somehow had the foresight to create these kind of networks. [Now] we are leading the world.' The beneficial effects of this foresight on the conduct of science was noted.

> NETWORK SCIENTIST: Because of the networks, across Canada we are doing science in a manner that I don't think could possibly have happened before ... A very large piece of the scientific community is [now] involved in promoting collaboration – inter-university and interdisciplinary, not just geographic. That is a very positive thing.

> NETWORK SCIENTIST: The network is like a national lab without the consequences – the bureaucracy, the nine-to-five mentality. Here, it's academic, competitive, but then we get together and we figure we're all part of this same process.

> NETWORK SCIENTIST: We created research groups that would not have existed otherwise, that spanned the country. Or involved different components of the country where we might not otherwise have encountered each other. These are cross-country collaborative interactions.

However, CGDN did not evolve in quite the way the program's architects envisaged. They had anticipated large-scale, cross-country collaborations. But for some reason – institutional logistics, egos, distance? – that did not happen. And, despite mutual goodwill, the number of researchers involved in one-to-one, bench-level collaborations was fewer than was hoped for.

NETWORK SCIENTIST: We don't interact on a project-by-project basis as much as was hoped we would. I think we fail a little bit there, just because there is too much to do and no time.

NETWORK SCIENTIST: There are some collaborative projects within the network. But it's not as heavily networked as it could be, I think.

NETWORK SCIENTIST: I have not been one of the ones who has interacted ... perhaps as much as some other people. Because I don't really have a collaborative project with anybody in the network ... it's not because I'm not interested, it simply hasn't been beneficial.

Still, by creating the intellectual and collegial infrastructure described above, the network allowed individuals to formulate different questions and approach their science differently. So, even in the absence of hands-on collaborations, researchers benefited from their interactions in the network: 'I don't think we would have done that project in quite the same way if it wasn't for the network,' one researcher reported. Another confirmed this view: 'We have changed in the way we ask questions and, therefore, the questions that we answer and what we publish. I know that for me – the kind of science I was doing, the directions I was taking – it's very, very clear that I do things differently than I would have done before.'

But because each phase of funding added new researchers, institutions, and industry partners, the capacity for collaboration and the nature of the network community were not static. The orientation changed over time.

Phase Transitions

When CGDN was renewed for Phase II, with its enhanced emphasis on commercial results, it meant more industry partners and more emphasis on commercial potential at the annual meetings.[7] Yet the overall ethos stayed much the same. Largely, this was because the core-set remained unchanged and because the expansion had been relatively modest: from twenty-one to thirty-three researchers, and from eleven to thirteen institutions. So the growth was easy to absorb. That was not the case in the transition from Phase II to Phase III. With the expansion to fifty researchers in more than twenty institutions, intimacy was almost impossible. Almost all of the network's founders felt

TABLE 4.1
CGDN growth in partner institutions and principal investigators, comparing Phase I and Phase III

	Phase I	Phase III
PRINCIPAL INVESTIGATORS	21	50
UNIVERSITY PARTNERS		
– Alberta	N	Y
– Calgary	Y	Y
– Laval	N	Y
– Manitoba	Y	Y
– McGill	Y	Y
– McMaster	N	Y
– Montreal	Y	Y
– Ottawa	Y	Y
– Queen's	Y	N
– Toronto	Y	Y
– UBC	Y	Y
– UVic	N	Y
TOTAL UNIVERSITIES	8	11
HOSPITALS AND INSTITUTE PARTNERS		
– Biotechnology Research Centre, UBC	Y	N
– Children's and Women's Health Centre, UBC	N	Y
– Children's Hospital of Eastern Ontario, Ottawa	Y	Y
– Hôpital Ste-Justine, Montreal	Y	Y
– Hôpital Saint François d'Assise, Laval	N	Y
– London Health Sciences Centre	N	Y
– Mount Sinai Hospital, Toronto	N	Y
– Hospital for Sick Children, Toronto	Y	Y
– Montreal Children's Hospital	N	Y
– Montreal General Hospital	N	Y
– Ottawa Hospital Research Institute	N	Y
– Robarts Research Institute, London	N	Y
– University Hospital, Vancouver	Y	N
TOTAL HOSPITALS AND INSTITUTES	5	11

that the culture changed radically at that point and that something important had been lost. As one commented, 'In the early days I knew everybody and now I don't. That happens when a group gets big enough. It means that we're now more of a conglomerate than a bunch of guys working together.' A comparison between Phase I and Phase III appears in table 4.1, showing the growth in the numbers of investigators and institutional partners.

Although the elite recruitment criteria applied in the first two phases caused a fair amount of debate, and many participants were uncomfortable with the emphasis on exclusivity, this approach produced a strong and cohesive culture. As a result, when the approach was reversed in Phase III, it tended to undermine what had been built to that point. One of the network's founders had spoken strongly in the past about including all qualified scientists, but when that eventually happened, he found the effects disturbing.

> We had such stringent criteria in the beginning and then, in order to get the Phase III funding, we had to open it up again. Wide open. That was a most difficult decision for me. I was not very happy about opening the thing wide because it was so indiscriminate. Some people were recruited just for their name. They didn't really have any interest in the community. They are part of the network and as yet I still haven't seen any contribution from these people.

Because so many people and institutions were now members, maintaining the same level of familiarity was impossible. The mechanisms of interaction that worked so well in a relatively small group stalled when numbers grew. People were disappointed that they could no longer get to know each other in the same way. Fear was expressed that a more corporate, commercially oriented style of doing things would undermine collegiality. Even the tenor of the scientific meetings – the great binding mechanism of the past – was affected.

> NETWORK SCIENTIST: The meetings haven't been great. All scientific talk; no play. This year's meeting was held in the middle of Vancouver, in a small hotel, where there was nothing that you could do together for fun. And it was tied to another huge conference. So everyone had been away from home too long, and [we] were too tired to play together.

> NETWORK SCIENTIST: It is immediately obvious when you go to a network meeting that this is not ... the style that we have been used to. These are meetings where the commercial aspect of what we work on is stressed. That's probably the biggest thing. And then the scientific content comes second.

Not only were the meetings different, the sense of commitment was different, too. When researchers were recruited for Phase I and Phase

II, it was for the long term. Renewal of funding was not guaranteed, of course; competition was fierce, and anxiety on that score was high. But no one sensed a finite horizon. In those early years, funding could be lost in only two ways: either the whole NCE experiment would be cancelled, in which case all the networks were in the same situation; or a network would not be renewed because its proposal would be judged inferior to others, and that was the luck of the draw. No third contingency, no sunset provision, appeared until Phase III. Many found it shocking and bitterly ironic that when the NCE program was made permanent, in 1997 – removing fears of overall cancellation – it was at the cost of continuity for individual networks. Thus, the researchers recruited for Phase III came in knowing that, at best, they would be with this group for a maximum of seven years. Together with the sheer numbers of new recruits, the sense of finitude limited the participants' 'buy-in.' In fact, by this point, several scientists were members of two or even three networks.

The relationships, and the willingness to trust, were no longer in place. Attitudes to the annual scientific meetings provide one example. In the past, attendance had been mandatory, not discretionary. But to many of the Phase III recruits, this was 'just another meeting'; they did not bother to attend. As one of the managers complained, 'The minimum that we ask is that you come to the annual scientific meeting. The old groups from Phase I and II are always there, but there is a much weaker understanding among the new recruits of why they need to be there. Some of them from the new group just didn't come.'

The funding bait was so diluted, because of the large number of researchers, that it no longer offered sufficient inducement. As well, many of the associations included in the Phase III proposal were included for strategic reasons. The purpose was to *simulate* dynamic expansion; actual connections were tenuous at best and in some cases divisive. For example, principal investigators had been recruited from Mount Sinai Hospital in Toronto, but historical disagreements marred relations between this team and their neighbours across the street at Sick Kids. The most recent disagreement concerned the administration of funds for Genome Canada, the umbrella body for genome research founded in 2000.

> Genome Canada is very much the legacy of CGDN. And we [Sick Kids] worked very, very hard to get the government to do that. And I think it is just a crying shame that we at this institution, the place where most of the

> genetic diseases work is done, are not being given the job of making sure the money goes to the right places. It is going to go to Mount Sinai. It has been diverted. There is a lot of political stuff that goes on. If Mount Sinai is going to use the money for genetic disease, that would be great. But it sounds like it is going to be diverted to doing all kinds of rubbish that has got nothing to do with genetic disease.

In a climate of tenuous connections and actual rivalry, the authority to compel attendance was lacking. As a result, enculturation into the network was minimal.

Network Science?

Grounded in laboratory practices and commercial motivations, molecular biology is an example of a 'practical science.' Divisions between the creation of knowledge (theory) and its applications (practice) are largely rejected. Meaning collapses into application, and truth value collapses into use and exchange values.[8] The focus is on converting laboratory results into profitable new therapies. What happens when individual research programs in molecular biology (medical genetics) are brought together under the banner of 'network science'?

Science is normally conducted in a highly competitive environment; individual labs are pitted against each other in races for resources and priority. At the same time, *within* a laboratory and under the direction of its leader, people cooperate, share resources and ideas, and publish together. In a sense, CGDN extended the boundaries of 'the laboratory' to include everyone (and every*thing*) in the network. All members of the network were considered colleagues; all had access to the network's technologies. In the long run, this extended definition of the laboratory proved 'more important to the scientific enterprise than a lot of the rest of what CGDN does, because this is where the new ideas and approaches that power everything else will be generated.'[9]

The ethos of trust and cooperation allowed network researchers to reduce competition. They helped each other with scientific problems, reviewed each other's papers, exchanged students, and advised each other at all levels. These tangible and intangible aspects of belonging made the network a coherent and cohesive entity. CGDN provided an organizational structure, albeit a loose one, that contributed to the production of first-class science. But whether this science could be described as a distinctive form of 'network science' is an open question.

Initial CGDN network and program documents contained descriptions of a clearly defined network research program, divided into projects and themes, with teams of researchers working together under the direction of project leaders. These descriptions might lead one to imagine the discussions at the start of each phase regarding what 'we' were going to do next and to imagine scientists working together, according to plan, to discover genes and therapies. But the reality of network science proved to be elusive. Network science did not exist where one might expect – in the 'network' research program – but it was very much in evidence in the services provided to members by core facilities and their directors. Understanding this development requires an understanding of the field of medical genetics.

Medical Genetics: An Overview

The science of CGDN is medical genetics, the field that studies the relationship between human genetic variation and diseases.[10] Genetic disorders are classified into three types: single-gene disorders, chromosomal disorders, and multifactorial disorders. Single-gene defects are caused by a single critical error in the genetic code. More than 4,000 single-gene disorders have been described. Chromosomal disorders are due to an excess or deficiency in the number of genes contained within an entire chromosome. The most common example is Down syndrome (trisomy 21).

Together with the effects of environment, multifactorial inheritance is responsible for a wide range of disorders that are believed to be due to multiple genetic mutations. Some cancers, coronary artery disease, and diabetes mellitus are included in this group. A mutation is defined as any permanent change in the nucleotide sequence of DNA. Mutations may occur in somatic or germline cells, but only germline mutations are inherited. Somatic mutations, however, are responsible for many medical problems. For this reason, cancer and coronary artery disease are often considered 'genetic' diseases.

The practical goal of medical geneticists is to understand the basis for mutations and to use that information to design new therapies for gene-related disorders. The field includes many rapidly advancing areas of interest, such as chromosomal analysis, cytogenetics, biochemical genetics, clinical genetics, population genetics, genetic epidemiology, developmental genetics, immunogenetics, genetic counselling, and foetal genetics.

Michael Hayden's research program in Huntington's disease is one example of the types of crossover that occur. Hayden's team has identified a marker used in genetic testing for Huntington's disease. As well as researching the genetic basis of the disease and testing for it in patients, they are also involved in pre-natal testing and in studying the psychological consequences of genetic testing on patients.

The history of medical genetics and the history of the gene are intertwined. Evelyn Fox Keller traces an arc through three periods.[11] The early twentieth century was dominated by a very powerful discourse of gene action. But the gene itself remained a statistical entity; a black-boxed construct. In general, medical science paid little attention. Interest increased when the physical basis of heredity was established, but mainly among those who studied rare anomalies. But little progress was made until 1953, when James Watson and Francis Crick described the molecular basis of DNA.

The mid-twentieth century was the era of early molecular biology, which seemed to provide answers to questions about the nature of the gene and gene action – the 'genetic program.' At this point, according to Barton Childs, medical genetics began in earnest, following the functional definition of 'one gene, one enzyme.'[12] In the 1960s, the development of the structural definition of the gene meant that inborn errors of metabolism could be described in terms of protein differences. The comparative youth of the field can be illustrated by network scientist Charles Scriver, who learned biochemical genetics when it was in its infancy. When Scriver joined the McGill University faculty in 1961, he was the first biochemical geneticist in Canada. He recalled that he was 'the first one formally trained to do that type of thing and be taken onboard as a person who would do biochemical genetics.'

In the late twentieth century, the molecular definition of the gene led to a technological explosion that moved genetic and molecular analysis beyond rare, single-gene disorders to complex, multifactorial diseases. The tools of molecular genetics underwent revolutionary changes following the discovery of a way to 'recombine,' or edit, DNA to permit the transfer of specific genetic traits from one micro-organism to another (see figure 4.1).

> The methods used in rDNA technology are fairly simple. We take, for example, the sentence (gene) for insulin production in humans and paste it into the DNA of *Escherichia coli,* a bacterium that inhabits the human digestive tract. The bacterial cells divide very rapidly making billions of

Figure 4.1: Restriction enzymes and their operation

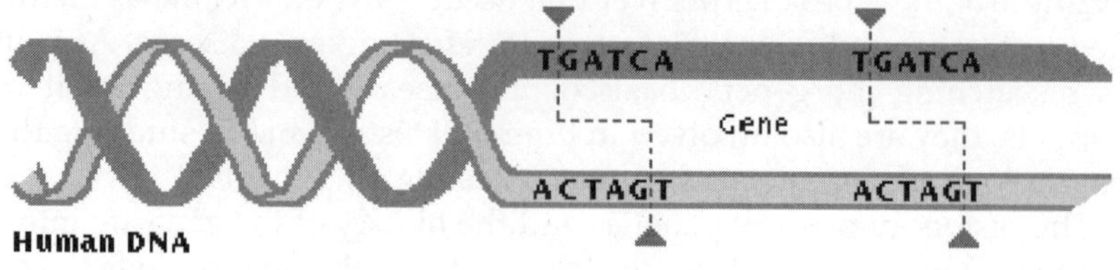

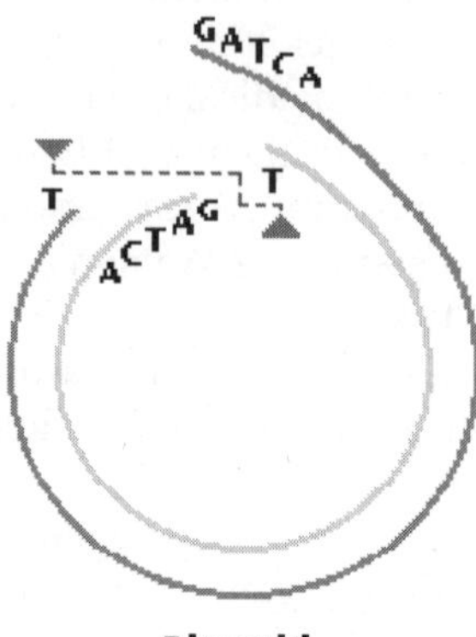

A **restriction enzyme** (▼) always cuts a sequence of nucleotides at the same site. In this example, the cut is made between Thymine and Guanine in the sequence TGATCA.

The same sequence may appear in the DNA of many organisms, or multiple times in one strand of DNA.

A **plasmid** is an additional, circular molecule of DNA found in almost every kind of bacteria.

Source: http://encarta.msn.com/media 461535878/Restriction Enzymes.html #rev (accessed 14 Sept. 2005)

> copies of themselves, and each bacterium carries in its DNA a faithful replica of the gene for insulin production. Each new *E. coli* cell has inherited the human insulin gene sentence.[13]

The editing analogy is appropriate, since molecular 'scissors' and 'glue' are used to 'cut' genes from the original DNA and 'paste' them into plasmids. The many varieties of restriction enzymes provide the scissors that precisely cut the specific base sequence of the DNA molecule. The 'glue' is an enzyme called DNA ligase. When the recombinant plasmid (the edited text) is inserted into a cell, the cell will process the instructions contained in it and pass along the instructions to the next generation of cells (gene cloning).

Minute amounts of DNA can be 'amplified' by the use of polymerase chain reaction (PCR), commonly thought of as a 'genetic photocopier.' The amplified material can then be fed through high-throughput sequencers and analysed by the sophisticated software of 'bioinformatics.' It is the marriage of biotechnologies and computer technologies that makes molecular tools so powerful. The availability of these tools, and the therapeutic promise of genetics, led to the founding of the Human Genome Project (HGP) in the late 1980s. As the century turned and the project neared completion, molecular biology again changed radically, as fields like proteomics and functional genomics came to the fore.

The 'new genetics' is revolutionizing medical genetics. It raises the prospect of altering the genome to *prevent* disease rather than *treat* disease. The ethical questions related to this prospect are troubling. Virtually all disease progresses as a combination of environment and genetics ('nature *and* nurture'). Medical geneticists believe that 'nature' plays the most significant role, and they act to intervene. Many believe this prospect raises the spectre of biological determinism and a new eugenics.[14] For others, the new genetics ignores the significance of 'nurture' – the socio-economic determinants of health and disease.[15] These debates and issues are compelling, but they lie beyond the scope of this study except where they have a direct impact on CGDN.

Space and Scale

In her comparison of high-energy physics and molecular biology, Karin Knorr-Cetina described the latter as small-scale, 'benchwork science' geared to 'treatment and intervention.'[16] By definition, molecular biology manipulates small objects in small laboratories. Its modest scale is such that a new laboratory can be assembled from the contents of two cardboard boxes: one containing a powerful PC, pre-loaded with genetic analysis software, and the other containing slides, reagents, and biological materials – a veritable 'Lab in a Box,' or 'Lab to Go.' The physical infrastructure of the laboratory is provided by the university, but the space and benches are generic. Beyond unpacking the boxes, nothing special is required.

The spaces of these laboratories and their architectural boundaries are standardized and predictable.[17] For example, the laboratories at the Centre for Molecular Medicine and Therapeutics (CMMT) are laid out in such a way that the upper and lower floors are virtually identical. A common room/kitchen is located on each floor, at one end of the hallway. This area is the social focus, with much coming and going. Signs on cupboard doors advertise meetings, seminars, and social events. Groups of graduate students and postdoctoral fellows chat over coffee and microwaved food at the common table. Overheard conversations: 'I had to sacrifice my first mouse last night'; 'I just found a mouse up my sleeve; its tail was sticking out. I thought I'd lost it.' (The mouse core facility was located at CMMT at this time.)

The laboratories are situated around the circumference of each floor, and the heavy and/or shared equipment is in the centre. Each lab appears to have two working benches in a bay, and a computer desk. The building's architecture provides no 'public face,' not even a functioning reception area. All exterior doors are locked and electronically controlled. None is identified as the main entrance to the building. The most likely candidate carries a sign advising visitors, in no uncertain terms, that they are at 'the wrong place': 'This is not the hospital,' it says. Those who persist must use the intercom to ask someone to come and physically admit them. Indifference to (or fear of?) public intrusion is a spatial feature of many, if not all, of the network facilities. The sites of knowledge production are not 'open.'

These sites, molecular biology labs, house 'biological machines' for the genetic engineering of knowledge. Karin Knorr-Cetina calls these machines 'prolific small-scale factories' for the mass-production of cell lines, bacteria, viral vectors, and the type of 'knock-out mice' discussed by the graduate students referred to earlier.[18] The network supplies these animals to order from its mouse core facility. Mouse models ('animal helpers') are research tools. Geneticists engineer them by 'knocking out' particular genes to try to cause disease. The mice are bred to be exactly the same; a blastocyst injection into the ovum changes the organism. These mice are not 'natural'; they are constructed in the laboratory. Bruno Latour discusses the 'purification' of wild nature that occurs in a laboratory.[19]

Comparison studies of the cultures of high-energy physics (HEP) and molecular biology show that experimentation in HEP involves large and very expensive experimental devices and hundreds of scientists. These huge investments demand a long-term communitarian orientation to the management of spaces and technologies. Thus 'big science' such as HEP is largely a collective enterprise. Publications list hundreds of authors in alphabetical order; discourse is open and free flowing along 'confidence pathways' that link people together; a variety of spokespersons represent the work. Knorr-Cetina calls this a 'post-traditional communitarian structure.'[20]

In contrast, molecular biology's 'Lab in a Box' has no dominating technical apparatus that would focus a community. Unlike HEP, it is highly individualistic, rather than collectivist, in orientation. For example, there is a tradition of naming laboratories after the leader (the Hayden lab; the Worton lab); leaders speak for and represent the lab as a whole. They are the focal point for public and scientific recognition. They appear in the media, give papers at conferences, and accept the awards, while those who actually do the work often go unrecognized.

Studies show that the most prominent and authoritative 'experts' are those who are farthest from bench research. A dual system is at work. Teams of postdoctoral fellows, graduate students, technicians, and junior faculty do the actual hands-on science under the direction of project leaders, while the lab director attracts the resources and plans the research program.[21] Lap Chee Tsui, chair of CGDN's Scientific Advisory Board and head of the International Human Genome Organization (HUGO), said, 'I'm still in the lab in terms of interactions but not day to day, not hands on anymore. I have to rely on people telling me what is going on. Of course I miss it. But it would be very difficult to go back. Because now I design experiments so complicated I need people to help me out.'

Given the dominance of laboratory leaders and the fragmentation of molecular biology, CGDN's achievements in fashioning 'something like' a communitarian network culture, and 'something like' network science, are worthy of comment. Unable or unwilling to overcome embedded epistemic norms, they were able nevertheless to scale up until the network approximated 'big science.'

Scaling Up

Until the 1980s, molecular biology in Canada was a competitive and fragmented world, where solitary researchers in small laboratories conducted small-scale experiments. Interactions were limited, at least partly because of the time and costs involved. As one of CGDN's investigators recalled,

> You might see your research colleagues at meetings or even make special trips to go to their lab and discuss research in common. And you might even send some grad students around or a technician to learn a procedure or something. But that was a relatively small number of interactions that each lab would have with another lab ... There was [no] money there. You could [not] justify saying, Well, I would like to go over and see so-and-so do this, [and] take it out of your operating expenses.

But as the research issues became more complex, it was increasingly clear that molecular biology could no longer operate effectively on a small scale and remain competitive internationally. By the time Michael Hayden reached out to colleagues across Canada in 1988, it was already unlikely that a medical geneticist, working alone, would find both the gene and subsequently the cure for a genetic disease. A more likely scenario would combine the work of medical geneticists and other molecular biologists with viral agents, tissues, genetic physicists, pharmaceutical chemists, gene-sequencing technologies, 'purified' mice, and bioinformatics.[22] Like high-energy physics, biology was becoming 'big science.' Lap Chee Tsui clearly described the differences:

> The way we do science is definitely different now than it was, say, fifteen years ago. Back then it was all very small experiments. And, of course, things were very primitive, too. Medical research has definitely changed – its scope, the way it approaches things, the knowledge required to run or operate it. It is no longer just a solitary person dreaming up some experiment. It definitely requires quite a lot of help from other people. And if not from other people, from computers and the Internet. Before, the literature and meetings were the only things we had. You got all your connections that way. Now the scope has just broadened so much.
>
> To undertake a biological question, you need engineers and statisticians to come in. A single person can't operate effectively in biology anymore. I don't know how to put it. Compare biology to physics. In physics these days, although a few are still doing investigator-driven research in small laboratories, seeking answers to a few very specific questions, the bulk of the experiments are done by big groups, large-scale networks using central facilities. I think biology is moving towards that model.

Through the NCE program, Canadian biologists were able to aspire to the benefits of big science. NCEs helped the Canadian life sciences earn respect and remain internationally competitive in medical genetics, protein engineering, bacterial diseases, neuroscience, respiratory diseases, and other biological areas. According to one of CGDN's founders, 'The network has been very good for the field of medical genetics in Canada. It has strengthened the discipline. People regard Canada as being a good place to do genetics.' Another network researcher compared his experience in CGDN with his experience in the United Kingdom.

> In England, I [belonged to] a large collection of scientists working on a similar topic. The group is so big it's like a force of nature. In that type of institute you are immersed in science in a way which we can't do in Canada. We don't have the resources. We can't allocate that much money to do focused research of that type. *But that's what we're doing here in the network. We're doing focused research* ... The network allows us to bring together a critical mass of people who think about medical genetics problems, from different perspectives. And I think that's a real strength.

A Network Research Program?

Although all the CGDN participants were involved in something to do with human genetics, this 'critical mass of people' was *not* focused at first. It took time to develop an understanding of what a *network* research program was and to weave together the projects of individual researchers in a coherent way.

When the founding researchers were recruited in 1988, they were asked to write up a 'wish list' of projects they would choose to undertake if funding were available. One researcher recalled that, when Ron Worton visited his lab to invite him to join the network, 'He said, "Well, have you got projects that you are not doing now but you would like to propose?" And I said, "Oh, yes. There are always lots of those."' The desiderata of individual researchers were then creatively combined to constitute the *network's* research program in the funding proposal.

Thus, the 'network research program' was an imaginary program, rhetorically constructed from *individual* research programs to obtain funding. What was proposed was simply a continuation and expansion of ongoing individual studies, with some of the expansion being due to network funding. The overall scientific objective of this composite was to study the molecular basis of genetic disease and the genetic basis for susceptibility to common diseases. The major goal, at that point, was to clone the genes responsible for selected genetic disorders. This would evolve in later phases, but in 1988 geneticists were still preoccupied with 'gene-hunting.'

Little changed when the network became operational. Early NCE assessments criticized the emphasis on the individual researcher: 'for the most part, [the science] seems to be too much PI-driven and not enough project-driven.'[23] Over the years, however, CGDN became more astute at shading annual reports and statistical materials to convey the *impression* of integrated research projects and active lab-to-lab collaborations, despite the relative paucity of both. As one manager admitted, 'We always said we had research projects, because that's what we were supposed to have, but we didn't, really. We had people working on different diseases ... So it was pretty hard for us, at the end of the day, just to describe what our projects were.'

In the original proposal, individual projects were loosely grouped under seven themed headings:

1 Identification of disease genes based on chromosome location; for example, cystic fibrosis, Huntington's disease, myotonic dystrophy, Wilson's disease
2 Mutation and functional analysis in Duchenne muscular dystrophy, retinoblastoma, and retinitis pigmentosa
3 Genetics and biochemistry of inborn errors of metabolism; for example, in Tay-Sachs disease and Sandhoff disease
4 Analysis of genetic factors predisposing to common diseases in mice and humans, using recombinant congenic strains in mouse models of human disease and amplified sequence polymorphisms
5 The structure of human genetic variation, such as thalassemia in French Canadians and Tay-Sachs in French Canadians and Ashkenazi Jews
6 Construction of chromosome specific cDNA maps for specific tissues, including retinal cDNA isolation and mapping and linkage analysis in diseases affecting the retina
7 Core technology facilities – nine technologies were offered in Phase I.

At the end of Phase I, this research program was assessed by an expert panel, which based its assessments on self-reports submitted by the network and a two-day site visit by the panel to the network's head office in September 1993.[24] Descriptions of 'themes,' 'projects,' and 'teams' were accepted at face value as part of an integrated program. The panel recommended trimming some projects, focusing others on more competitive fields of research, and regrouping physicians and scientists into smaller numbers of highly competitive teams.[25] But overall, in the panel's estimation, CGDN had achieved 'outstanding progress.' If there were an international standard in genetic research, it said, the network 'might well be on top of such an international comparison.'[26] The panel submitted a favourable report to the NCE directorate on 25 October 1993. In part, that report read,

> The Site Visit Committee noted the outstanding role played by scientists in this network on the international level with respect to the cloning of disease genes and investigating their functions ... The Committee was also impressed by the collegiality and networking established among the investigators of the network and noted the importance of the establishment of the core facilities as a catalyst in this process. The Site Visit Committee, therefore, enthusiastically recommends that the network continue.[27]

On 28 October 1993, three days after the expert panel had submitted its favourable report, CGDN tendered its proposal for Phase II of the NCE program. Although it built on what went before, it restructured the research program to accommodate the research interests of new recruits. The research emphasis would now switch to common multi-gene disorders such as Alzheimer's disease and breast cancer instead of the rare single-gene disorders that had been the focus of Phase I. According to Ron Worton, this was a pragmatic decision, made because 'if we don't get into the complex diseases, the reviewers are going to wonder why, and they're not going to give us funding for Phase II.' Even more pragmatic was the fact that these were *profitable* diseases. As another researcher commented, 'The big pharmaceutical companies are interested in these big polygenic diseases ... diabetes, inflammatory bowel disease, the sort of things that tens of thousands of people suffer from. Because that is where they are going to make their money.'

The Phase II research program included eight themes:

1 Identification of disease-causing genes
2 Genes and phenotypes
3 Dynamic mutations (novel causes of human genetic disease)

4 Genetic analysis of complex traits (mouse models of human disease)
5 Genetic epidemiology and population genetics
6 Therapeutic interventions for genetic diseases (new theme)
7 Applications of molecular genetics to health care (new theme)
8 Core facilities

The two new themes emerged from the new emphasis on *relevance* in program criteria, which weighted translation of findings into practice equally with excellence of fundamental research. Theme 6 was a move into gene-based therapeutics and clinical trials; theme 7 was a foray into commercial diagnostics.

By the end of Phase II, CGDN had adopted in its reporting a language of 'key discoveries,' 'breakthroughs,' and 'commercial impacts.' It maintained metrics on all, claiming 170 discoveries overall in Phase II, of which 100 were related to common, multigene disorders. Twenty 'key discoveries' were highlighted, including the isolation of the first two Alzheimer familial disease genes by a researcher at the University of Toronto in 1996.

The discoverer was new to the network that year; he had been recruited when he was close to the breakthrough after working on the project for a number of years. Even though he allocated only 10 per cent of his time to the network, CGDN was able to claim credit because he was a member at the time the genes were cloned. Another of the new Phase II researchers identified breast and ovarian cancer mutations in the genes BRCA1 and BRCA2. These too were claimed as 'network discoveries.' On the other hand, it was one of the original PIs – a 1988 'young researcher' who had spent almost his entire career with the network – who discovered a family of proteins that inhibit cell death. This breakthrough was quickly patented and spun out into a company (see chapter 6).

In the new theme concerned with therapeutic interventions for genetic diseases, researchers had not yet translated findings into applications; rather, they had 'created tools for gene-based therapeutics, setting the stage for therapeutic advances in Phase 3.'[28] Progress had been made in biological problems in haematology that had been barriers to the use of gene therapy for blood diseases and in the use of herpes simplex virus (HSV) as a vector for gene delivery. The second new theme, applications of molecular genetics to health care, demonstrated much more translational progress. For example, headway had been made towards the identification of a direct genetic marker for osteoporosis risk, based on estrogen receptor variants, and of predisposing genes for risk of coronary artery disease (atherosclerosis). In addition, one of the researchers developed a novel technology for the rapid, accurate, and cost-effective DNA sequencing of mutations that was quickly adopted by the Human Genome Project. Also, key advances had occurred in the mutation analysis of the gene for retinoblastoma (Rb), a devastating childhood cancer of the eye. Because each Rb mutation was revealed as virtually unique, efficient methods for mutation analysis were required. This need was translated by the researcher into mutation diagnostic reagents and kits for cost-efficient diagnosis and cascade testing in families. The investigator noted that, without the network,

> we might never have developed the Rb test the way we have. We would have failed, like every other lab in North America, to practically help patients, because the test would have been too expensive, and too difficult. [Without the network] I don't know where I could have got funding to do that research.

The network submitted its progress report on Phase II, together with an application for Phase III funding, on 29 April 1997. In February 1997, the NCE program had been made permanent, but individual networks – including CGDN – had been 'sunsetted.' At that point, the Phase III funding proposal had been in preparation for almost a year. In less that two months, it had to be reoriented towards achieving sustainability after the cessation of NCE funding. The research program was collapsed into the four elements with the most potential for commercial exploitation: identification of disease-causing genes, pathogenesis and functional genomics, genetic therapies, and genetics and health care. A two-day site visit was arranged for late June.

The panel concluded that funding should continue for the maximum allowable period: until 31 March 2005, subject to mid-term review in 2001. It cited the increasing number of multiple-authored papers across projects as an indicator that 'the group now shows much more evidence of working together as a team,' and concluded that the network's evolution had been nothing short of 'remarkable, in that it has not only achieved its stated goals in fulfilling the mandate established for NCEs, but in almost all cases has surpassed them.'[29]

CGDN's weaving together of individual strands to give the appearance of coherence, such that reviewers were convinced the network had 'achieved and surpassed' its stated objectives, was a considerable rhetorical achievement. But whether the credit belonged to the network or the individual researchers is an open question. It remains unclear how much of the *network's* research program would have been achieved if the network had not existed, and it is difficult if not impossible to calculate the incremental value the organization added to existing individual research programs.

Recognizing these ambiguities, CGDN revised its organizational purpose. Until early 2001, its mission had been 'to research the diagnosis and treatment of genetic diseases and to help move the resulting discoveries into the health care system.'[30] Subsequently, the network defined itself, more accurately, as an 'enabling organization' and a 'catalyst' for research.[31]

Core Facilities: 'Where Spokes Converge'

In one aspect of CGDN's research program – the provision of core facilities – there is little doubt about CGDN's contribution to the researchers' accomplishments. Core facilities are the advanced technologies and technological expertise that helped network investigators speed research progress and breakthroughs. They were the enabling technologies on which the network's research program rested, and they legitimate the network's claims and justify the notion of 'network science.' In the opinion of one researcher, 'The core facilities are a kind of network legacy ... They are really the axle where a lot of the spokes converge.'

In defining an NCE, the network metaphor itself is less useful than is the concept of 'spokes' and 'hubs.' This is also the case with core facilities. Network researchers across Canada (spokes) drew on core facilities and expertise (hubs). The hubs supplied the network's material and intellectual infrastructure. Rather than proceeding from researcher to researcher, collaboration was between researchers and core facility directors – the network's 'master collaborators.' The nine core facilities and directors available in Phase I appear in table 4.2.

CGDN's core facilities simulated the technological support infrastructure of 'big science.' Easy access to powerful and expensive technologies allowed relatively small laboratories to undertake ambitious projects and to compete internationally. In a priority race to identify genes, a race in which every additional day matters and specialized technologies may not be available at a researcher's home university,

TABLE 4.2
CGDN's core facilities, Phase I, end of year one (1990–1)

Facility	Purpose	Director(s)	Institutions
Computing and genotyping	Analyzing population genetics	Morgan	McGill
Cell bank	Long-standing research collection	Scriver	McGill
In situ chromosome hybridization	Radioactive detection of short probes	Duncan	Queen's
Protein analysis	Improved sequencing, reagents, protocols	Aebersold	UBC
Transgenic and knockout mice	Animal models	Jirik and Marth	UBC
Hybridoma	Production of monoclonal antibodies	Lee	UBC
Electron microscopy	Service provision	Lea	U of T
DNA sequencing	Sequencing genes in small fragments of DNA	Tsui	U of T / HSC
Somatic cell mapping	Mapping cells to specific chromosomes	Worton	U of T / HSC

Sources: CGDN-AR (1991); CGDN-EP (1993); CGDN-FP (1988)

the core facilities enabled resources to be dedicated to a particular project and to move it ahead rapidly. The network would fund core facilities when it could balance demand and supply – that is, when demand for a novel and/or sophisticated 'leading edge' technology could be matched to a principal investigator who was ready to act in the capacity of director and willing to offer that technology to other members of the network.

The status of the core facilities developed at the end of Phase II and the beginning of Phase III is shown in table 4.3. By this point, three DNA sequencing facilities were supported: a new, large-scale sequencing site at the University of British Columbia, a small fragments core at the University of Victoria, and the original in Toronto.

By the time of the Phase III mid-term review (May 2001), CGDN had shifted its emphasis in a major way: the network's new mission was to be a catalyst for research advances in the wake of the sequencing of the human genome. 'We are now in the post-genomics age. Many genes involved with pathology have been cloned. The focus now shifts to the proteome and pathogenic mechanisms.'[32] The core facilities were rationalized into four clusters:

TABLE 4.3
CGDN's core facilities, end of Phase II, beginning of Phase III (1998)

Facility	Purpose	Director(s)	Institutions
Bioinformatics	Service and training in computational biology	Oulette	UBC / CMMT
Complex traits analysis	Screening congenic mice to identify control genes	Skamene	McGill / MGH
Fluorescent in situ hybridization	Mapping genes and cDNA to chromosomal regions of mouse and human genomes	Squire	U of T / OCI
DNA sequencing	Large-scale sequencing	Hayden	UBC / CMMT
	Original facility: small fragments	Scherer	U of T / HSC
	Small fragments	Koop	UVic
Genome alteration in *C. elegans*	Comparative genomics: nematodes and humans	Culotti	U of T/ Mount Sinai
Genome alteration in mice	Knockout mice	Rudnicki	McMaster
		Jirik	UBC / CMMT
		Nagy and Rossant	U of T/ Mount Sinai
Core computing and genotyping	Original population genetics database	Hudson and Morgan	McGill / MGH
Immunoprobes	Developing reagents for cell and molecular experimentation	Wilkins	Manitoba
In vivo DNA analysis	Mapping DNA-protein interactions	Drouin	Laval / SFA
Protein-protein interactions	Isolating and identifying interacting proteins	Friesen, Greenblatt, Pawson	U of T / B and B
Transcribed sequence detection	Analysis of genome sequences for gene discovery	Rommens	U of T / HSC

Sources: CGDN-AR (1999); CGDN-EP (1997); CGDN-FP (1997)

1 Core technology platforms: DNA sequence analysis, bioinformatics
2 Gene technologies: in vivo DNA analysis, genotyping, transcribed sequence detection
3 Protein technologies: immunoprobes, proteomics
4 Genome alteration: *c. elegans*, mouse

As before, the highest demand was for DNA sequence analysis. A partial cost-recovery program shifted some of the burden for facilities maintenance from the network to the users, reflecting the Phase III focus on sustainability. All the participants interviewed agreed that the core facilities, and the skills of their directors, represented one of the network's key legacies:

> RESEARCHER: The core facilities were a real catalyst for promoting interactions. We did a lot of cross-country running about among different labs, but a lot of them centred around core facility usage.

> RESEARCHER: I think a key feature of the network has been the [core] facilities, especially the sequencing facility. There is no way I could have got that sequencing done without the resources of the network.

> RESEARCHER: For me, the high point of the network has been the core facilities. That's been my favourite component of the network. It's been fantastic.

> CORE FACILITY DIRECTOR: If you want something immediately there is immediate cooperation. When we know that someone is getting close to a gene, and they need this kind of help, we put the secondary requests aside and emphasize this competitive project.

> NCE SELECTION COMMITTEE: The committee attributed the success of this network to an exemplary collegial exchange of knowledge and its reliance on and extensive sharing of resources, such as the core facilities. Genetic research, especially human genetics, is extremely costly to perform. The committee considered that the sharing of core facilities alone represents a significant benefit from the investment.[33]

The added value was in setting up an infrastructure for undertaking the technical work that no single researcher could afford to set up independently, in his or her own laboratories, but needed to use sporadically. Gene mapping was an example of this advantage. When the original core facility was set up, a backlog of demand quickly accumulated. As the director recalled, 'If somebody wanted something mapped, they just sent it to me, and it was a given that I was going to do it ... If they hadn't been part of the network, they would have had to organize for just one little probe to be mapped with somebody else.'

As sophisticated and expensive technologies like this became more and more central to research progress and as demand for them increased, universities and hospitals started to acquire their own capacity. At that point, network resources were redirected to other technologies not yet generally available. Core facilities would also be terminated if they were not used enough. For example, as can be gathered from tables 4.2 and 4.3, seven of the ten Phase I facilities had been replaced by the first year of Phase III (1998–9), when CGDN offered eleven core technologies in fifteen locations. In between, other core facilities had been started and abandoned.

In CGDN's first decade, approximately 20 per cent of the network's total program funding was dedicated to core facilities. The system appears to have been a cost-effective way of sharing resources. A number of researchers argued that, to encourage collaboration, *all* the network's resources should be directed into the development and maintenance of these platforms rather than into the relatively inconsequential amounts of funding allocated to each PI. Core facilities directors sometimes dedicated most or all of their individual network funding towards the support of their platforms: 'We throw our network money to the core facility – two people and about 600 mice ... and various equipment and instruments.'

Theoretical Perspectives on Core Facilities

The existence of core facilities meant more than just sharing expensive equipment and biological materials; it also resulted in the pooling and sharing of expertise. They were an efficient way to leverage the productivity of researchers and ongoing research. Rather than duplicating facilities at different sites, they concentrated resources at one site *and in one person*. As a network researcher explained, 'It is the expertise of the people that is core, rather than the machines.' In fact, it is the *combination* of people and machines that counts, according to another informant: 'The core resource is one thing, and the experience of the director ... and the people who work there, is another.' The combining of machines and their directors in this way constituted what Bruno Latour calls a human/non-human hybrid and Andrew Pickering describes as a human-machine interface.[34] Such 'cyborgs' can find answers far more expeditiously than any 'regular' scientist or technician would. The issue is familiarity and the way constant practice refines skills. As a PI explains, 'I don't want my technician to have to learn a whole technique to do ten samples. That is a waste of everyone's time and money, quite apart from the machine.'

As technical and scientific experts, core facilities directors operated at the intersection of the network's material culture and moral economy. The material culture of a science is its 'tools of the trade': the machinery and methods of knowledge production, its instruments and experimental practices. The moral economy is the social rules and customs that regulate access to the material culture, establish authority over research agendas, and allocate credit. Robert Kohler pointed out that 'tools and methods only become productive when they are part of a social system for socializing recruits, identifying doable and productive problems, mobilizing resources, and spreading the word of achievements.'[35] The interesting question for Kohler is how material culture and moral economy operate together to make research productive. Pickering argues that the mechanism is the 'mangling together' of human agency and performative material devices in 'a dialectic of resistance and accommodation.'[36]

These perspectives relate to the following combination of factors:

- the researcher's requirement to have results processed (e.g., genes to be sequenced)
- the budgetary resources required to mobilize machines and/or technical staff to do the processing
- the power of these machines and technicians to produce inscriptions and standardizations from the data supplied
- the technical and scientific expertise of the core facility director, who manipulates the technologies, even when they resist, to process the experiments

When machines, money, molecules, and magi are related in this way, what becomes apparent are modest, 'local' actor-networks of human and non-human elements that become nodes in the larger actor-network that is CGDN. From their location at the nexus of science and technology and knowledge and expertise, core facilities are as much a form of 'know-how' as of fundamental 'know-that.'

Core facilities directors are 'master collaborators' because, by virtue of their position at a hub, they are aware of and participate in the majority of research projects and can suggest potentially fruitful interactions between researchers who previously may have been unaware of each other's work. One director observed that 'in the early days ... I was among a small number of people who were actually connected to most other people in the network ... Virtually everybody had been storing up a bunch of stuff that they wanted mapped ... I interacted with a lot of people.' More than that, however, directors could actually steer the direction of a project and the research agenda.

> By virtue of running a core facility, I know a lot of things that are going on, like new projects and stuff. And I have had input ability to actually participate and to help steer some of the research. A researcher will come to me and say that they want to do something and I say, well, maybe it wouldn't be good to do it that way, it's better if you do it this way. You see what I mean? I can actually play a role in determining the projects. If you're in a core facility, well then, everybody is coming to you and saying, 'I want to do this, what do you think?' And so you have a chance for having input there.

In terms of the communal life and the moral economy of science, and the rules of mutual obligation that prevail in these communities, Robert Kohler argues for the centrality of three particular elements: *access* to the material culture; *equity* in assigning credit for achievements; and *authority* in setting research agendas and in deciding what is worth doing.[37] Under this definition, it seems clear that a substantial portion of the network's moral economy relies on the central role of core facilities directors.

Summary: The Durability of the Ephemeral

This chapter has presented two contradictory impressions of CGDN. On the one hand, there is a sense of the chimerical, an ephemeral community with an 'imaginary' research program: now you see it, now you don't. On the other hand, there is a sense of real durability: established relationships founded on mutual trust and anchored in significant technologies. Is the black box empty or full?

The question can be approached by looking at the shift between phases. The addition of new actors into an existing network is always destabilizing. New actors come with their own networks, each of which has its own goals. Stability requires the disconnection of alternative associations so that the network becomes the only point of passage. A process of mutual shaping must take place to incorporate the new into the existing actor-network. That integration was successful in the shift from Phase I to Phase II. But in Phase III, the enrolment of new allies (researchers and institutions) seems to have taken place without enough attention to *interessement*. The latter occurs where network builders *lock in* potential allies by gaining their commitment to a set of goals and a course of action. *Enrolment* without *interessement* creates a fragile network that readily fragments. The Phase III expansion was overwhelmingly strategic, and as a result alignments were incomplete and the voice of the *spokesperson* no longer spoke for all. Under conditions like these, the system's stability becomes precarious: black boxes open, points of passage are ignored, and ambivalence becomes pervasive.

What, then, to make of *strong* associations that only seem to strengthen with time? Perhaps it is useful to think of networks within networks; layers of associations like tree rings, showing different stages of expansion. The older layers are the most dense; compacted; difficult to *dis*associate. The newer layers are more porous; they can be peeled apart and peeled away. As well, it is clear that materiality makes networks durable and that more durable materials tend to produce more stable networks. Ideas and talk are ephemeral; as John Law argues, to persist they need to be embodied in inanimate materials such as machines, books, and buildings.[38] The core facilities thus 'anchor' the network in complex and costly technological tools and in the embodied knowledge of the scientists and technicians who operate them.

5 From Science to Commerce

NCEs were funded with the intention that they would, among other benefits, generate products and technologies for profit. Although 'excellence of the research' was the dominant criterion in Phase I selection and remained a background condition, commercialization and partnerships with the private sector were critical to the core mandate. As the sunset of NCE funding loomed, CGDN focused on constructing a portfolio of licensing deals and spin-off companies that they hoped would provide a stream of future revenues. *All* alternative sources of income were investigated. The goal was to prime the 'pipeline' that links the laboratory and the market, the process of traversing which has been called 'arduous, passionate, rich in ritual, and steeped in conflict and controversy.'[1]

Two major strategic shifts occurred in the network's relation to 'the pipe' and its evolution 'from science to commerce.' The goal of the first, in the mid-1990s, was to bring some coherence to the commercial portfolio. The second, in the late-1990s, was motivated by the desire to sustain the network beyond withdrawal of program funding. The initiative ratcheted networking to a higher level; it brought the three original life-sciences NCEs together, to combine pipelines and jointly finance, 'bundle,' and market the resulting portfolio of technologies.

Understanding the Pipe

The pipeline metaphor originates in the linear understanding of innovation that underpinned the postwar social contract for science (see figure 1.1). Even proponents of the 'open science' model now view the linear model as an unrealistic depiction of the public/private, basic/

applied relationship, especially in forefront sciences such as information technology and molecular biology, which 'overflow' attempts to contain them. Yet the pipeline metaphor survived the collapse of the linear model; it remains ubiquitous in the talk of molecular biologists, as well as in the policy discourse. It is, in fact, 'the spontaneous philosophy of scientists' and has been used in public debate since the end of the nineteenth century.[2]

Certainly, the metaphor accurately represents the realities of commercial development in the life sciences where a ten- to twelve-year evolutionary process links the discoverer's laboratory bench with the packaged, brand-name product on the pharmacist's shelf. As figure 5.1 shows, for the process to begin, patents must first be secured on a gene or pathway discovered in the researcher's laboratory. The patents are then licensed out, sometimes to the researcher's own university-based spin-off company, sometimes to an independent biotechnology company. The patents and licences attract the venture capital required to fund 'translational research' (this concept will reappear in chapter 6) – the process of proving out the discovery and seeing if it will scale up for commercial development. If scaling up and early clinical trials are successful, smaller biotechnology companies will often merge in order to bundle their candidate technologies and advance them further. Eventually, a partnership will be entered into with a pharmaceutical company large enough to command sufficient resources to navigate the late-stage clinical trials and regulatory approval process.

The length and complexity of the pipeline made the NCE program's expectations of commercial prospects unrealistic. The federal government had a poor understanding of how long it takes to move 'raw science' out into the market. Its attitude, said a policy adviser, 'was short-termist and linear, very linear: "We will do some research, we will have a result and we will make a product and we will sell it."' As a senior CGDN scientist commented about anxieties on this score,

> We were really very scared that it would be impossible to get renewed if they expected us to produce a line of products and a group of connections in five years ... It's taken us into the third term to begin to produce what they thought we were supposed to do from the outset. Which was to create the links with the private sector, to produce the spin-off companies, to generate patents and products. And I think that's just about the right timeframe. Ten to twelve years is the realistic timeframe.

Figure 5.1: The drug discovery and development pipeline

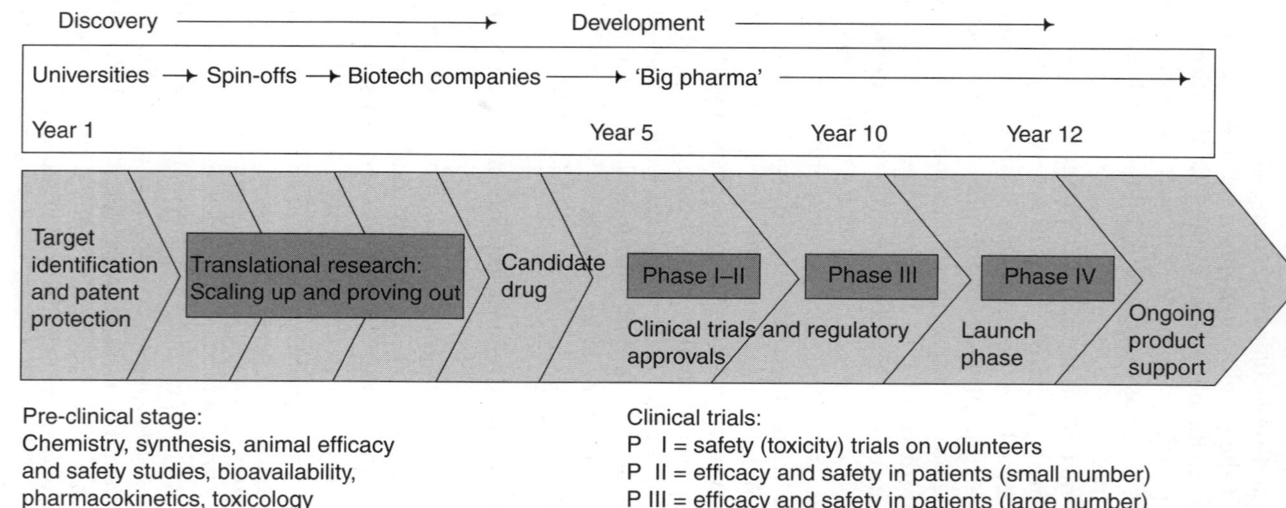

Pre-clinical stage:
Chemistry, synthesis, animal efficacy and safety studies, bioavailability, pharmacokinetics, toxicology

Clinical trials:
P I = safety (toxicity) trials on volunteers
P II = efficacy and safety in patients (small number)
P III = efficacy and safety in patients (large number)
P IV = effects in actual use

Contributing sources: Dr Campbell Wilson, Astra Zeneca; Dr Jim Brown, Glaxo Smith Kline; Dr Usher Fleising, University of Calgary

There is no shortage of good ideas; good ideas are plentiful. But it takes a great deal of time, money,[3] and effort to steer a discovery from the front end of the pipe through myriad competing ideas to commercial success at the far end of the pipe. Although 'ideas are cheap,' most do not survive. 'For every hundred academics that spot something they think is commercially interesting,' said the network's CEO, 'only one will actually get it together to carry it through to the marketplace. The other ninety-nine ideas just languish.' This attrition rate was one reason for the concern about the government's expectations. Even the pharmaceutical industry was disturbed about federal misunderstandings of the way the pipe worked. As one of CGDN's industry partners stated in the network's first annual report,

> It is important to realize ... what the time frame is likely to be for the emergence of product candidates, especially in the pharmaceutical area. It is important that this [network] research be government funded, *and that renewal of funding not depend on the commercialization of products in academic research centres.* This is the best way to assure that academic research stays at the cutting edge in each field, and generates the unexpected discoveries that can be pursued and developed in strong industrial research centres.[4]

This is an ardent defence of the division of labour in the linear 'open science' model: government funds science; science publishes results; industry takes up and develops results. Under a strategic science regime, however, the state hopes universities and research networks will generate revenues by patenting and commercializing their own discoveries. This expectation interferes with the traditional division of labour and increases transaction costs for industry.

Because of the risks and costs involved in commercialization, network and university technology managers hedge their exposure by maintaining portfolios of discoveries 'in the pipe,' each at a different stage of translation and financing. (These are intimately related; new financing is needed to fund each succeeding phase.) In CGDN, Phase III activities focused on the far end of the pipe, as the strategic plan moved from translation to speculation; that is, from the early-stage scaling up of research results to speculation in finance and investment vehicles and venture capital funds.

Industry Partnerships

Industry partnerships are not as extensive as a cursory perusal of program or network documents might indicate. Many alliances are listed, but most involve minimal commitment and funding. A willingness to sign on to the formal network agreement is an indicator of who is and who is not a 'real' industry partner. In CGDN's case, only two private-sector partners signed the first formal network agreement: MDS Health Ventures Inc. and Merck Frosst Canada Inc., and only Merck Frosst made a funding commitment – $70,000 a year for three years to provide research fellowships. A third 'industrial partner' signatory – BR Centre Ltd – was actually UBC's Biotechnology Research Centre, where three of the network's researchers worked. (This body was also listed as an institutional member.) Naming the centre an industrial partner was a fiction that helped disguise the fact that little attention had been paid to the NCE mandate for industrial linkages. Michael Hayden and Ron Worton simply imported their respective long-standing relationships with Merck Frosst and MDS into the network. As one of the founding group commented,

> We didn't know what industry partnerships meant. We needed partners and the government kept saying the partners must contribute in a direct fashion. But, obviously, there had to be a desire on the part of industry to participate and some means for them to feel that this is worth their time and effort. They weren't going to join us to make charitable contributions. There was also the concept of in-kind [contributions], which was, in those days, very primitive. We didn't know what in-kind really meant. So this was an extremely difficult thing for us to cope with ... nobody knew what the rules should be and nobody knew what the government was looking for.

The Phase I funding proposal also listed Pharmacia (Canada) Inc., Squibb Canada Inc., and an entity called EuGENE Scientific Inc. (a strange name, given the flawed history of eugenics) as potential partners.[5] The majority of discussion in the section of the proposal on 'Potential for New Products and Processes for Commercial Exploitation' related to EuGENE. The company was to be the network's research and development corporation, a public-private joint venture between the Hospital for Sick Children at the University of Toronto and MDS Health Ventures Inc. Half a dozen pages were given over to EuGENE's prospects, products, capitalization, and profile in the network. According to the proposal,

> EuGENE Scientific is a new company which is determining its goals as a direct consequence of the proposed establishment of this network ... It is

> primarily because of the proposed involvement of network investigators that MDS laboratories has agreed in principle to make an investment in the order of $2 to $2.5M to this company.[6]
>
> Scientists who are part of the network will participate as scientific advisors to EuGENE for development of their gene probes for diagnostic tests [and] diagnostic kits ... The scientists in the Network see the establishment of EuGENE as vital to the proper exploitation of their gene probes.[7]

However, EuGENE proved to be short-lived. Between proposal and legal agreement, the company changed its name to Kronem Systems, then quietly disappeared, apparently despite investments from NRC-IRAP and the MDS investment fund. No further trace of the company could be found. The industry liaison office at the Hospital for Sick Children believed that it ceased operations in 1991 or 1992.[8]

Another phantom company haunted the proposal for Phase II funding, submitted in 1993. The industrial linkages section of that proposal was structured around a spin-off called NGI (Network Genetics Inc.), which had been formed to commercialize network research. The language of justification on diagnostics and therapeutics was similar to that used for EuGENE:

> In an effort to create Canadian receptor capacity for CGDN's intellectual property, the network has taken the bold step of launching a new venture [NCE Genetics Inc. or NGI], the first Canadian company focused on genetic diagnostics and therapeutics. This is part of a long-term strategy by the network to capture value in Canada and enhance Canadian commercial contributions in this area.[9]
>
> The ultimate competitive edge for this company is based on its special relationship with network researchers [which] represents an invaluable source of commercial and market intelligence which will assist in ensuring the development of new IP ... The new network venture will begin its commercial activity within the next months and start the process of technology transfer.[10]

According to the Phase II proposal, NGI had hired a scientific director. Its financial and business plans would be ready by the end of the year; and the proposal confidently predicted that the company would be operational in early 1994. But, as with EuGENE, after the renewal award, further references to NGI ceased.

Subsequently, according to network documents, during Phase II CGDN researchers developed understandings with twenty-one authentic companies or corporate divisions, of which three were network spin-offs (see table 5.1). Of the $10.2 million generated from these connections, almost two-thirds ($6.4 million) came from two 'big pharmas': Merck Frosst ($3.4 million) and Schering Canada ($3 million). Most of the Merck contribution related to their support for the new Centre for Molecular Medicine and Therapeutics at the University of British Columbia, while Schering's investment is for the presenilin genes project (Alzheimer's disease).

The nature of these relationships, and the degree to which they were attributable to network facilitation, is not clear from the documentation. The network classifies them as 'industry collaborations' for reporting purposes but also refers to them as 'sponsored research.'[11] Apart from the Merck relationship, the majority of these linkages appear to be arrangements whereby network researchers are funded to further develop patented technologies licensed by the company. Where the relationship is with a spin-off company, the amounts reported parallel the funds raised in the investment community to advance the patented technologies.

Despite what an observer might conclude from program and network discourse, little evidence exists of *bench-level* collaborations between academy and industry researchers, working together to advance technologies along the pipe. A CGDN private-sector board member confirmed that, to the best of his knowledge, 'there are no network-private sector collaborations in the same sense as there are network-public sector collaborations based on the relationships amongst the scientists.' The main factor inhibiting bench collaborations is that industry labs are largely concerned with product development, whereas researchers are in the business of knowledge creation.

Industry rarely involves itself in collaborative basic or even translational research. With the possible exception of Michael Hayden's work with Merck Frosst on Huntington's disease, CGDN researchers cited no examples of working directly with researchers in industry. According to a policy analyst, this was the case for the NCE program in general:

> Side-by-side bench collaborations are few and far between. I can't think of an example offhand. I think that it's rare. Collaboration is [defined] much more in terms of planning and monitoring the research and dealing with

TABLE 5.1
CGDN-industry relationships, Phase II

Company	Cash Invt. (C$ 000s)	Principal Investigators	Institution	Project
Amgen	130	Dick	Sick Kids/UT	Stem cell technology
Apotex	132	Gallie	Ontario Cancer Institute/UT	Retinoblastoma protein
Apoptogen (spin-off)	1,194	Korneluk, MacKenzie	CHEO/UOttawa	Apoptosis/cancer
BioChem Pharma/Gene Chem	328	Skamene, Gros, Rouleau	Montreal General Hospital/McGill	BCG therapy, bladder cancer, congenic mice
Connaught	130	Morgan, Skamene	Montreal General Hospital/McGill	TB/BCG genotyping
Glaxo-Wellcome	25	Hayden	CMMT/UBC	Huntington's disease treatments
IBEX Technologies	70	Scriver	Montreal Children's Hospital/McGill	PK treatments
ID Biomedical	15	Jirik	CMMT/UBC	Genetic testing technology
INEX Pharmaceuticals Inc.	287	Cullis	UBC	Liposome carrier therapy
		Worton	Ottawa General Hospital/UOttawa	
		Tsui, Dick	Sick Kids/UT	
Leo Laboratories	50	Rousseau	St Francis Hospital/Laval	Psoriasis
MDS-SCIEX	370	Dovichi	University of Alberta	DNA sequencing technology
Merck Frosst	793	Triggs-Raine	University of Manitoba	Yeast 2 hybrid/tyrosine phosphatases
		Jirik	CMMT/UBC	
Merck Frosst/CMMT	2,448	Hayden, Jirik, Hieter	CMMT/UBC	CMMT
Merck, Sharpe, Dohme	151	MacLennan	Banting and Best/UT	Phospholamban interactions
Millennium	48	Gros	Montreal General Hospital/McGill	Cloning LPS locus
Myriad Genetics	60	Rommens	Sick Kids/UT	Breast cancer
NeuroVir (spin-off)	405	Tufaro	UBC	Neurological/HSV gene therapy
Rhone Poulenc-Rorer	446	Hayden	CMMT/UBC	Lipoprotein lipase therapy
Schering Canada	3,000	Hyslop	Centre for Research in Neurodegenerative Diseases/UT	Presenilin genes/Alzheimer's
Visible Genetics	30	Gallie	Ontario Cancer Institute/UT	Retinoblastoma/malignant hyperthermia
		MacLennan	Banting and Best/UT	
Xenon BioResearch (spin-off)	80	Hayden	CMMT/UBC	Gene identification in unique populations

Source: CGDN-FP (1997: 20).

disclosures and IP issues and training and so forth. I can't offhand think of an example where two people actually sat side by side at the bench and did things.

Large pharmaceutical companies tend to wait until small biotechnology start-ups and spin-offs have completed early-stage proof-of-concept and development work – then they buy the company. Their unwillingness to collaborate at more basic levels of the pipeline understandably causes a degree of resentment among researchers.

RESEARCHER: If you are looking for a disease gene, forget it. Nobody is going to support you in terms of a company, a commercial business. You want to isolate genes for diabetes? They say 'good luck.' But if you already *have* a gene, then, yeah, they are very interested. But the support doesn't come until you have a gene. You have to have a result.

RESEARCHER: The big ones, the Pfizers and the Glaxos of this world, they haven't been anywhere near the network. Despite lots and lots of overtures to try and get them to show some interest ... They are not interested in big collaborative projects with basic scientists at all. They want to do clinical trials and they want to do basic research in their own facilities where nobody can see what they are doing and they get all the patents. They don't want to be involved with basic researchers in universities ...

RESEARCHER: They will come and pick stuff up. If they see you doing something interesting that they like, they will come and try and pick it off you. But they don't want to work with you on it ... If you look at the stuff that they *are* funding, a lot of it is clinical trials. So what they are basically doing is they are getting the government to help them do their clinical trials. I mean they are laughing all the way to the bank. I am very cynical about this. I have been at it a long time and I have watched this stuff and I have tried to talk to them about doing some basic stuff and they don't want to do it.

Even Michael Hayden, an indefatigable booster of industry and a close collaborator with Merck Frosst, admitted that support from big pharma was weak. Acknowledging industry's legitimate right to serve their shareholders, he nevertheless argued that the private sector has underinvested in fundamental research while overinvesting in marketing. 'That's my only real criticism,' he said. 'Too much has gone into marketing, and that doesn't really help us, doesn't fund students, doesn't fund postdocs.' Industry needed to be reminded that, without basic research, there would be no products to market.

Realizing that hands-on partnerships with big pharma were unlikely to happen and that it had to meet the technology transfer mandate of the program, CGDN needed a commercialization strategy. After failing to develop one in house (see remarks above regarding EuGENE Inc. and NGI Inc.), the network turned to the private sector and hired in the talent it needed to move its intellectual property into and along the front end of the pipe.

Traversing the Pipe

Because scientific excellence was the dominant criterion in Phase I, no one in Ottawa or the networks paid much attention, at first, to how the other mandate points (relevance to industry, linkages and networking, management) would be implemented. Neither federal bureaucrats nor network scientists really knew what was required or what the ground rules were. For example, it wasn't until 1995 that the federal position was clarified on whether networks could hold equity in spin-off companies. As a result, opportunities were lost because holding shares in start-up companies was an alien concept for most academics. 'Equity was almost like a dirty word in those days,' recalled Michael Hayden. Although CGDN assisted in the incorporation of Visible Genetics Inc. (VGI), now a multimillion-dollar public company, it received no shares in return. Network managers often cited VGI as a lost opportunity and always listed it as a network spin-off. However, the PI most associated with the company suggested that this would be stretching the truth. 'They count it. And I don't mind them counting it, but really, they had nothing to do with it.'

Attitudes began to change in Phase II, when the weighting of the selection criteria changed, putting commercial development and industry partnerships on a par with research excellence. In order to be renewed for Phase II, networks had to adopt a much more aggressive stance on commercialization and industry linkages. A policy adviser described the change:

> The networks said just stand out of our way, step back, we are coming through. And this is especially the case in the medical networks. They had huge amounts of money on the line and they said, just stand out of

> the way. We don't want to hear about policy or programs. Forget about the ILOs, they don't know what they are doing. Universities, step out of it. It's ours. We need to sign a deal ... tomorrow and it is going to be for hundreds of millions. Just get out of the way. So it really took on a life of its own.

In Phase II, CGDN changed its research direction to one that was potentially much more profitable. In Phase I, the focus had been on simple, single-gene diseases, such as cystic fibrosis and Tay-Sachs, which are potentially lethal but also relatively rare. These diseases are of little interest to pharmaceutical companies because not enough of the population is affected by them for drug development to be commercially viable. On the other hand, complex, multigene diseases – cancer, heart disease, diabetes – are much more common. They affect a large percentage of the population and are therefore potentially much more profitable.

The First Commercial Turn: Profitable Diseases

Although the idea of turning away from pure, curiosity-driven research towards 'profitable diseases' made some researchers uncomfortable, the network's Phase II funding proposal presented a research program that, while still directed towards basic understanding of the phenomena, took a much more hands-on approach to commercial applications and partnerships. This was the first 'commercial turn.'

Arie Rip speaks of 'promise requirement cycles' and the pressures for credibility that result when 'promises ... become an accepted means of exchange between scientists and sponsors.'[12] Many promises were made in the Phase II funding proposal about translating network discoveries into commercial applications. When the network was renewed, those promises had to be met.

Principals describe a sense of controlled panic as they struggled, and failed, to come to grips with the implications of their commitments. Recognizing their limits, they went outside the academy and recruited the network's first commercial director. That individual came in with an impressive background in the biopharmaceutical industry, a one-year contract (subsequently extended to a second year), and a large compensation package. In her words, 'They paid me a lot more than would be typical for that kind of position. Plus, I had a bonus. A big bonus ... In fact they paid me more than the managing director. They just decided that I was what they needed.'

Her mandate was to try to organize the network's scattered portfolio of projects into something that made business sense. She found a group who knew they needed to 'do' commercialization but had no idea how to go about it. 'There was zero experience,' she recalled. 'I mean zero. When I arrived, they were making a company by pushing all sorts of things together that didn't go together. It just didn't make sense. None of them really knew how to get into it.'

CGDN was desperate for someone to bring commercial order to the chaos. A collective sense of relief took hold as the new director moved in. She visited all the network nodes and PIs, working to understand their projects, calculating what could be bundled together to create spin-off companies and what might be better as pharmaceutical company collaborations. She asked for, and received, a pool of money – about half a million dollars – to invest in projects that needed a little more time in gestation before anything commercial could be attempted. This became the strategic fund, which remained operational for the duration of the network and under which scientists could apply for $50,000–$75,000 to investigate the commercial viability of their projects. By the end of my fieldwork in 2001, the strategic fund had advanced some $800,000 to network investigators who were looking for commercial relevance in their research.[13]

The network gave the new director a free hand in the use of her financial resources, time, and focus. She became the entrepreneur-in-residence, teaching scientists about 'what makes a company, what makes a product, and how you put those things together.' After reviewing the situation, the director identified about half a dozen solid, commercial opportunities that could be realized quickly. From that first top-down assessment emerged two new companies (Neurovir Inc. and Apoptogen Inc.), the Alzheimer's project (Schering Canada), and a handful of patents. Essentially, the director elevated the network's commercial ambitions to the biopharmaceutical standard and taught researchers to 'look behind the science' for profitable opportunities. 'The problem is that the scientists see exciting science as opposed to seeing a product ... I was looking for products, not for science. To some extent I couldn't care less about science. So I was asking a different set of questions. I was adding a commercial rigour that wasn't there before.'

To validate her recommendations, the director was able to turn to the network's private-sector partners and board members for support. CGDN's industry partnerships were not extensive, and small start-ups and local biotechnology companies predominated. Nevertheless, the quality and commitment of individual participants was high. They were active at the annual meetings, reviewed network research, provided advice, and validated what the commercial director was doing. 'It meant I wasn't alone. It gave me a sounding board and support from ... the pharma side. If it had just been the scientists, they wouldn't have trusted my judgment.'

Less than two years after arriving, the commercial director moved on, but not away. In September 1996, she became chief operating officer of Neurovir, one of the two new network spin-offs she had helped launch. In retrospect, she felt fortunate to have arrived at the network when she did. The commercial opportunities that had been absent in Phase I were just beginning to emerge, and she was able to capture them. But maintaining a constant flow through the pipeline is a major challenge. And the opportunities in the pipe have to be substantial. Minor initiatives are rarely worth pursuing and certainly cannot support a company. When she left, little of magnitude was lined up in the pipe: 'I didn't see a whole lot more to go capture, which is one of the reasons I was ready to move on.' A senior member of the professional staff confirmed this recollection: 'In those days, there was not a lot to keep a commercial person fully engaged. The direction that we're taking now belies that to some extent. But back then, when she looked back down the pipe, she didn't see anything else coming up.'

The managing director added the commercial role to his responsibilities, and another staff member was recruited to help spread the administrative load. The network was able to consolidate its resources for a few months. Then, in the February 1997 federal budget, came the welcome news that the NCE program had been made permanent – CGDN had played a leading role in rallying the forces lobbying for its continuation – and the announcement of funding and criteria for the Phase III competition. At the same time, however, the government released a bombshell that few had predicted: a sunset provision. Networks would receive no more than fourteen years' funding in total; after that, they were on their own. Thus, it wasn't until Phase III, when the fourteen-year cap was announced, that science-for-profit became a survival priority for networks. This was the second commercial turn.

The Second Commercial Turn: Science for Profit

None of the researchers, managers, and private-sector partners I interviewed approved of the federal exit strategy. All felt that it was fundamentally misguided, given the length of time needed to develop commercial viability and taking into account the lack of receptor capacity and venture capital in Canada. Bureaucrats, they concluded, simply did not understand the way science works. As one senior policy adviser and NCE board member explained,

> Some of the people in the NCE [directorate] thought that [at sunset] what the NCEs were supporting, the fields of research, were then 'finished' because they would have put all their ideas into applied research. Which is a pile of BS because fields that are important keep moving, and you've got to stay with them if you're going to stay hot. The people doing policy don't have any experience [of the way science really works], which is really what the issue is. [A person like me] becomes a broker or an irritant [to tell them how science works].

Nevertheless, the NCE directorate's exit strategy *was* implemented and the network set about dealing with its ramifications. In the words of Michael Hayden, 'We've taken our destiny into our own hands. We don't really trust the federal government, or the provincial government, or our universities to secure our future. We're now saying we've got to do it ourselves.' The sunset provision meant that the Phase III proposal, which outlined the strategic plan for the years 1998–2005, had to lay the foundation for the network's sustainability beyond the end of federal funding.

The strategic plan for the Phase III proposal, which had been in preparation since the autumn of 1996, now had to be modified in light of the policy change. In assessing which existing projects and PIs should be continued or abandoned and which new PIs should be recruited, attributes of personal entrepreneurship and commercial potential came to the fore. Thus, researchers were assessed in terms of their ability to initiate and innovate; the numbers of IP disclosures, patents, and licences they had been involved in; their industrial collaborations, industrial consulting, and commercial advising; the amount of industrial funding they had raised; and the new diagnostic and therapeutic products or services they were developing. For any particular project in commercial development, hard questions were asked. Had partners or sponsors actually committed funding to the project? What were the stage of development and the funding ratio? And how coherent was the completion and exploitation strategy?[14] As one of the managers commented, 'We really had to get serious about the business of making money ... If the money wasn't there, we couldn't do science. So we had to hustle.'

The net was cast wide in an attempt to enrol commercially inclined researchers. If the proposal was approved, the number of PIs would almost double as new centres and programs were added. While network members endorsed the expansion, many did so reluctantly. As was discussed in chapter 4, the changes that were introduced in Phase III fundamentally altered the culture of the network. Until then, a balance had been maintained between traditional academic norms and new commercial values. Now, commerce came to the fore and collegiality suffered. Many of the founding PIs thought the changes were distasteful and distanced themselves. As one put it, 'The emphasis on profit as what we have to do makes me feel uncomfortable ... I regret the intrusion of profit and its more ugly form, greed, upon academic culture. I consider that a loss.' Another was upset that network funding to PIs became so contingent on market values: 'Basically the network is now in the mode where ... they are particularly looking for stuff which is commercial. It has got to have a commercial application, or else!'

The revised proposal was submitted in May 1997. In September 1997, CGDN was notified that it would be renewed for Phase III. A series of changes followed quickly. The founding managing director resigned effective February 1998. The network was subsequently incorporated as CGDN Inc., commencing the first fiscal year of Phase III as a not-for-profit corporate body. And, to complete the transition, a new chief executive officer was appointed with a mandate to make the network self-supporting by 2005.

Recruited on the basis of his past experience in forming companies and taking them to market, the CEO began his term in June 1998. From the start, his approach was strongly oriented to profit and to the rapid achievement of his mandate objectives. His first change was to the organizational structure, to the 'ephemeral informality' discussed in the previous chapter. The flat, collegial arrangements that had been in place for network management were replaced with a hierarchical corporate framework, one that reflected industry standards rather than those of the academy. His own title – chief executive – was an indication of the new orientation. As one of the professional staff commented a few months into the CEO's appointment, 'He is not a network man; he's not an NCE man. He's not going to have those collegial kind of relationships because his mandate's different, and because where he's come from is different. And his style is different. He's much more of a lone player.'

If the network was to aggressively commercialize its technology and interface with the private sector, the CEO believed, it needed a corporate facade. It should be recognizable by industry as a formal entity. Incorporation had been considered in the past but rejected. Now, the CEO wanted the network to have the practical ability to sign contracts and hire its own employees. He wanted an organization that was transparent to the finance and investment sector, and an organization that was disciplined by its structure.

> Eastern philosophies will say that with structure comes freedom. And I believe that. So to me, it's a natural extension for people, after a while, to realize that having a corporate structure will simplify a lot of relationships. Incorporation in no way restricts, in fact I would argue it enhances, the ability of the members to interact more freely, in a true network sense, because you're not always wondering what exactly the framework is that you're dealing with.

With the new structure in place, the CEO recruited as his commercial director a young lawyer-scientist with a background in negotiating biotechnology deals. A new strategy was developed and approved by network members at a planning meeting in June 1999, one year into the CEO's term. At the same meeting, PIs agreed to give up 10 per cent of their research budgets, if required, to enhance strategic activities.

The strategic plan involved three innovations. The first was an aggressive emphasis on commercial development. Scientists had to be motivated to recognize the commercial potential of their discoveries if the pipeline was to be kept primed. Finding likely prospects for that development was the task of the new commercial director. 'Probably ninety per cent of the best of the best of the best scientists in the country belong to this network,' she said. 'If *they* can't come up with technologies that can feed the pipeline, then I don't know who can.' In the network's earlier phases, the tendency had been to attempt to build companies around genes. That approach hadn't worked particularly well: its legacy was a clutch of small companies, built around single technologies, only a couple of which could stand on their own. The new goal was to look for platform technologies with broad applications and scientists who would 'stay the course':

> It has to provide a solution to a gap in the market. It has to have more than just the potential of one product. If it is a service, it has to be a service that is lacking or missing in the industry. You have got to have a scientist who is going to be entrepreneurial and stick with you along the way. They have to be committed. And the business plan has to make sense. There are lots of things, but the primary things are the technology and the commitment of the scientist.

The stance was proactive. Instead of waiting for scientists to call, the approach was to 'do rounds,' visiting every laboratory at least once a year. This helped the commercial director understand what was in development and allowed her to suggest a commercial spin where appropriate. Another strategy was to solicit for commercial prospects at the annual scientific meeting. 'Just being there and being in their face was an important reminder to the scientists that there was a commercial aspect to their work.'

Once the projects' commercial prospects had been identified, the CEO built partnerships with the finance sector to exploit them. 'He's a broker between the scientists and the money folks,' said a pharmaceutical industry partner. 'It's something that he does quite easily.' The CEO ratcheted up the commercial profile of the network by forming a for-profit company. This would allow greater scope for creating revenue opportunities and set up the network to partner with venture capital firms and receive a larger share of the profits. In the past the network had only been a facilitator; now it would learn how to be a full partner in financing the commercialization process.

But this could not take place without the cooperation of the other institutional partners: the universities and hospitals who owned the intellectual property. The CEO was unimpressed with the commercial abilities of public institutions. The public system is not, by definition, an entrepreneurial system: 'It was set-up for the mail to be delivered, for civil and criminal justice to be administered. It's not based on innovation, inventiveness, entrepreneurialism. It provides some basic infrastructure to allow society to do the day-to-day things.' As a result, asking the public system to act entrepreneurially was misguided.

The justification for transferring technologies from the public to the private domain and allowing venture capitalists and entrepreneurs to realize high returns was their willingness not only to pledge risk capital but also to manage the complexity of translating publicly funded technologies to the marketplace. How else would university technologies be converted into therapies and find their way to patients? The CEO entertained no scruples about privatizing public knowledge.

> People who think there's too much commercialization have never been involved in it. They have no idea of the difficulty in managing the hundreds of steps that all have to be successfully implemented in order to get that commercialization.
>
> [Without] the commercialization of scientific findings ... we'd still be suffering from plagues.
>
> Sometimes you commercialize a piece of physics which later turns out to allow laser operations on cataracts ... So, how can you arbitrarily say you're not going to commercialize technology?

Entrepreneurs do not draw boxes between public and private, between science and technology, he said, and neither should society. Public funding of basic research leads to cures for devastating diseases, 'but we also need a recognition that the process requires an entrepreneurial component to result in cures or therapies.'

The need to fill the pipe with public discoveries is far greater in a knowledge-based economy than it was in a resource-based economy. 'Where before you had prospectors that went around the north, picking up rocks, and looking for gold, you've now got prospectors going around universities, picking up ideas, and looking for knowledge.' Because the network had a track record of having done excellent science and of having put some of that science into the pipe, the CEO felt he could forge partnerships with these prospectors – venture capitalists – where the universities had failed. Shortly after his appointment, the CEO visited all the network's academic partners:

> We sat down with all the universities and medical institutions. We explained that all of the money we derive for research purposes flows out to researchers at their institutions. So in 2005, when government funding ends, if the network disappears, their researchers are going to see a negative impact. Then we explained what we're going to do to ensure that the network continues ... And once they understood what we were doing and why we were doing it, they were very supportive.

The second component of the CEO's sustainability strategy was a plan for building a $20 million endowment fund through a federally registered charitable foundation. 'We have a noble cause,' he said. 'We are seeking to find cures for a variety of gene-based diseases that inflict a lot of suffering and cost on society. So we can take that and we can build a foundation around that.' A retired investment banker, a board member, was asked to head up the effort, and the foundation appointed its own executive director. The model was a similar entity established by Neuroscience, a non-renewed network, that had established various funds and entities in an attempt to ensure its continuation.

Although willing to invest energy in the project, the retired banker was dubious about the prospects of raising enough funding to ensure the network's survival. 'We don't have the same pizzazz in fund-raising as the Children's Hospital or the Cancer Society or the Alzheimer Society,' he noted. 'We touch all of those diseases and many more, but somehow genetic research is not something that you can put your hands around.' The most recent figures available at the end of field-work (fiscal 2001) showed the foundation with a book value of $6.2 million, of which $1.3 million was cash, with the balance represented by estimated equity in CGDN start-ups.[15] With the subsequent loss of market interest in the life sciences and the unstable global situation, book value has likely declined. Even if achieved, a $20 million endowment would produce a maximum of $1.2 million in annual revenues, a figure that represents only 25 per cent of the $4.5 million in annual federal funding that would be lost.

So the network needed another activity to generate ongoing revenues, and this venture was the third, most challenging, and most controversial leg of the sustainability strategy. Traditionally, the network had earned an interest in intellectual property by helping researchers develop spin-off companies and by finding partners to commercialize network intellectual property. Now, through its for-profit company, the network would venture directly into the world of finance and investment. Not only that, it would do so in partnership with the other two 1989 life-sciences networks, which also faced the end of funding. The CEO's innovation was to get the boards of the three original life-sciences networks to agree to form their own venture capital fund – Excella Ventures – and seek investors jointly.

An investment fund could provide the profits to assist all three towards sustainability. The self-funding entity would converge on profitable therapies for a disease from three different starting points – genetics (CGDN), protein engineering (PENCE), and bacteriology (CBDN). This joint approach offered the promise of generating a critical mass of intellectual property – the kind of critical mass that would eventually grab the attention of big investors and big pharma.

The logistics were somewhat Byzantine: each of the networks had incorporated a for-profit subsidiary. These three subsidiaries were equal shareholders in a federally incorporated entity, Excella Ventures Inc., which, under CGDN's leadership, managed the development of the Excella Life Sciences Equities Fund. In the first instance, the goal was to raise $60 million, which would be invested in second-stage financing for fifteen companies over three years. This was the niche identified for the fund, since early-stage financing – the first $500,000 to $1 million – was fairly readily available in Canada. As the CEO explained,

> It's very difficult to generate that next round of money, the second-stage financing of five to ten million dollars. That's the market that we feel we can add significant value in. The reluctance to invest in that second round is because people know that for every hundred companies that get seed financing, ten, at max, are going to survive. And so, how do you pick the ten that have a hope of surviving through the next round? Most people do not have the sophisticated knowledge base to do that. We, at the networks, do. I mean if we can't evaluate the potential of [these] technologies, who can? There likely isn't anybody else in the world that can do as good a job as we can, because if we don't have the world leader in the field, somebody in our networks will know who the world leader is.

Although comprehensive and ambitious in scope, the CEO's three-tier strategy met with scepticism and resistance both within the network community and outside it.

Resistance

Several of the network's more senior researchers actively disliked the corporate facade. As one of them observed, 'That positive feature of being able to meet with other people and be part of a scientific society, it's just not the same when it's a corporation.' The new emphasis on profit was viewed with suspicion. In the words of a long-term member, 'It isn't what the scientists want. It isn't how we live or think or want to see things go forward.' Another stated, 'I never dreamed that we would be creating a foundation. To this day, I'm not sure that I agree with it. Or with the creation of a venture capital group. It never occurred to me that we would be sponsoring the creation of a venture capital group. And with the other networks, too.'

Some researchers felt that the new goals had little to do with providing health for Canadians, or advancing the research frontier, and everything to do with stock market speculation and commercial 'bubbles.' The resistance was not to commercialization as such, but to the *type* of commercial activity, which was beyond the comfort zone of many. The following observation is typical of the discomfort expressed:

> They were looking to fund the five most fresh, great, fantastic ideas as strategic business opportunities. You know, big and fancy and harebrained and virtual with little potential for contributing to health. That is where they were going to put all the [network's] money. And they were turning down translational research opportunities that would be of immediate and obvious impact! A simple little thing that actually turns a profit, that actually does a job and delivers health care, they didn't want to bother with.
>
> I got really mad because of it. To me the network's responsibility was *exactly* to do something practical and useful and now. And not go off into funding harebrained, high-risk stuff. To me, the best way to ensure [sustainability] is the safe way. You have to have some safe, solid things that are going to be solid, real companies. Old-fashioned ones that do something and pay dividends. Not ones that just get sold on the stock market and make money in a phoney way. That is really scary and that is not how we should build the strength of our network.

Some informed external observers detected a certain amount of hubris in the network's plans, especially in regard to Excella. As one private-sector funder commented, 'The networks are in danger of falling into the trap of forcing the scientists to raise money for its own sake, as opposed to using money to create something of lasting value.' One of the problems, according to the same observer, was that the networks were trying to do everything at once: 'They are trying to become commercialization vehicles, as well as science vehicles, as well as management vehicles, and that's doing too much. You can't do all of that.' Although networks had strengths in identifying candidate technologies and in managing the arrangements to get them commercialized, they were not going to be able to do it all, nor were they going to be able to retain all the value. 'And if they think that by starting their own little captive fund they are going create a lot of value, when today the pharmaceutical companies don't even have enough money to do it, that seems a bit ridiculous.'

In summary, pressure to become self-supporting changed the operational strategies of CGDN. Many in the network were uncomfortable with the new direction. A policy analyst noted that 'people have been worried about commercialization for a long time, but I think we are starting to see some of the dark side of that come out. It is very clear that there is a downside to it.' Others worried that the focus on commercialization would crowd out fundamental research. Nevertheless, the CEO was convinced of his path to self-sufficiency and enjoyed the support of his board and scientific director. The latter described the strategy as creating 'a legacy that is totally independent of government.' But the desire for 'total independence' did not stand up to scrutiny.

A Third Turn: Back to Basic(s)

In 2000, PENCE, a more commercially aggressive network than CGDN, analysed the income it could expect to generate from spin-off companies, based on existing performance, and concluded that self-sufficiency was not achievable by 2005, if at all.[16] According to a PENCE informant, 'There's just no way these companies have the financial capacity to fund the fundamental research, because they are not big enough. It's as straightforward as that.' Without fundamental research, the networks would have difficulty retaining their integrity as independent entities. Increasing focus on commercial goals was already distorting the accumulation of knowledge on which the whole edifice rested. 'And if you don't have a good fundamental research base you really don't develop any effective strategic or applied research programs.' This same conclusion was reached almost six decades earlier, in Vannevar Bush's report: 'Applied research invariably drives out pure. The moral is clear: it is pure research which deserves and requires special protection.'[17]

As a solution, the Excella concept met with a lukewarm response from universities and ILOs, as well as the finance community.[18] Even CGDN's two partner networks were less than enthusistic. One representative believed that 'the [Excella] fund is sort of a daydream ... but it won't work. Investors aren't going to put their money into something which hasn't been proved out.' So although Excella continued,[19] the

CEO's focus on speculation and short-term profit started to look passé. As a group, NCEs initiated planning on two broader alliances: broader both in terms of the number of networks involved and in terms of the underlying approach to knowledge development.

The first was a task force of board chairs working to develop scenarios under which government might continue to provide core research and training funding to successful networks beyond fourteen years, rather than have them shut down.[20] This group argued that if it made sense to fund a network in the first place, as a feature of strategic science policy (see chapter 1), then it makes sense to continue funding networks as long as they advance program goals. Desirable policy outcomes cannot be expected to continue once the state withdraws. As Arie Rip has observed, 'Scientists pick up the resources and run with them, and only in exceptional cases will they continue with newly initiated research lines after the funding stops.'[21] Although the program's director appeared open to discussion on the issue, it seemed unlikely that the decision would be reversed. A likely scenario is that, post 2005, successful but expired networks will reinvent themselves and reapply, as Ron Worton did with his successful bid for a Stem Cell Genomics and Therapeutics NCE, in 2001.

The second broad alliance was initiated by Dr Fraser Mustard, head of the Founders' Network, chair of PENCE, founder of CIAR, founder of PRECARN, architect of the Ontario Centres of Excellence, and prime mover of the NCE concept. It shows his characteristic flair for bold policy measures. Although in the very early planning stages at the time of writing, this or a similar initiative could prove to be an authentic reconciliation of public and private interests in the Canadian biosciences.

As described above, the chances that any one network could provide for itself from commercial revenues was highly unlikely. Even to attempt such a thing would be to seriously compromise the research endeavour. In Fraser Mustard's estimation, it was tantamount to 'expecting Galileo to make a profit.' He believed the commercial approach of the NCE program was misconceived from the start. It had not formed part of his initial conceptualization.

> We had proposed national networks in the *fundamental* sciences. At that stage, it wasn't conceived to be directly linked to industry ... But each time [the bureaucrats] kept making it tougher and tougher in terms of those commercialization requirements, because I guess that's how they were marketing it with the powers that be in the public service. Eventually, the

> program became so focused on private-sector involvement that, in a way, we were simply doing applied and strategic research with and for companies. And we were basically undermining our fundamental research base.

Even in the United States, Mustard argued, industry does not fund the biosciences; that mandate belongs to the National Institutes of Health. Inevitably, once a program becomes dependent on private-sector financing, it is steered towards an applied science mode and 'probably ends up being driven by huge pharmaceutical interests, which does not really give you the base for your fundamental research.' Without intervention, he argued, this would be the trajectory followed by NCEs as they approached the fourteen-year sunset.

To counter the trend, both a division and a concentration of labour were needed. In the first instance, NCEs and industry should each do what they do best: basic research and development, respectively. A similar conclusion was reached in Sweden, where the 'Research 2000' plan recommended that universities withdraw from the role of making *direct* contributions to industry. This position is essentially grounded in the 'open science' model. In the second instance, instead of having each network (and each university, for that matter) maintain its own commercialization directorates, the tasks should be centralized. It would make more economic sense to *jointly* undertake the tasks of commercialization. One organization could develop and manage an intellectual property portfolio; expand relations with venture capitalists and potential licensees; and, importantly, support proof-of-concept work. More beneficial still, a joint venture such as this would facilitate the 'bundling' of technologies. Technology managers interviewed for this study, whether in universities or NCEs, all mentioned the 'one-product, one-company' phenomenon. They acknowledged that bundling would solve the proliferation of unsustainable start-ups, but they did not know how to achieve that solution.

Although Excella was a start in this direction, Mustard sensed that it was situated too far down the pipe. As well, the profit orientation would get in the way of what he wanted to achieve, which was *pre-competitive* funding for proof-of-concept work. 'It is one thing to build a network of talent and produce ideas,' he said, 'but how do you know the ideas will work on a larger scale? Venture capital won't fund that, because there is no profit in it.' What was needed was something *between* basic research and commercial development, a cooperative (rather than for-profit) body, that would fund translation and scaling up. Mustard drew inspiration from the PRECARN model he had helped establish in 1988. PRECARN (the Pre-Competitive Advanced Research Network) is an industry-led, Ottawa-based consortium of thirty-nine companies.

PRECARN companies work together to develop receptor capacity for advanced basic research in robotics and artificial intelligence. The fundamental science is drawn from IRIS (Institute for Research in Intelligent Systems), by far the largest and most complex NCE. At the time of writing, IRIS was one of only two NCEs not hosted by a university or hospital. PRECARN hosted IRIS and provided a pre-competitive platform for member companies. Mustard proposed that a Canadian Network for Biotechnology Commercialization could, in similar fashion, provide a pre-competitive platform for the life-sciences NCEs. Proponents believe that, without such an entity, the world-class research knowledge developed in NCEs will continue to be licensed to foreign companies. Despite the 'Benefit to Canada' clause in network agreements, Canada is losing the benefit of its investment in the program.

Mustard and his allies were proposing a government-industry partnership. Government funding was seen as essential to offset the risk aspect of investing in life-sciences research. Lobbying by the initiative's powerful supporters caught the attention of Ottawa and the corporate sector. The proposal was presented to a meeting of NCE boards of directors, receiving a mixed reception. ILOs were also dubious:[22] a centralized vehicle for commercialization would represent yet another threat to their single-institution focus.

But once launched on a path, Mustard has rarely been deflected. Few are more skilled at moving issues onto the policy agenda. He is committed to finding a more rational, long-term approach to NCE sustainability and to renewing the vision of the program's architects: to build globally competitive *fundamental* research networks and the receptor capacity to exploit their results.

Summary: Marketization – From Public to Private

CGDN initiated a number of changes in its trajectory from Phase I through Phase III as a response to program demands for commercial relevance. More than any other factor, however, the federal policy decision to limit the length of NCE funding to fourteen years shifted the attention of CGDN officers from 'science' to 'commerce.' Ambitions progressed from relatively small-scale licensing and start-up activities near the front end of the pipe to large-scale speculative ventures much farther along, involving investment funds and 'high finance.' This latter move upset the delicate balance in this network between public and private interests, and between basic and applied science.

Privatization is generally considered in terms of the transfer of assets from one sector to another. The boundaries seem fairly clear: 'public' knowledge on the one hand, large corporate entities on the other. The image of cartoon capitalists might come to mind: portly moguls clutching bags of money while pillaging the commons. But the privatization that takes place in NCEs – and in universities, too – is far more nuanced. In most cases, the move from 'science' to 'commerce' occurs internally. A component of the public transforms itself into a component of the private to commercially exploit a discovery. A patent transforms some piece of public knowledge into a private commodity. A public research organization disappears behind a corporate facade. Solutions to disease become a locus of profit.

Beyond this nesting of public and private is the triumph of neoliberal attempts to 'marketize' the public sector. There are no cartoon capitalists here; if capital is the enemy, the enemy is within. The move has caused some disquiet in policy circles. The NCE program's architect has intervened to try to protect the basic science component of these networks. Success in this endeavour would return NCEs from the 'overflow' model to the 'open science' model. Whether the intervention will succeed remains to be seen.

6 Adventures in the Nature of Trade

Previous chapters have focused on the network as an entity, with an organizational history and culture and a mission to exploit technologies commercially. But without *scientists,* none of this would exist. Scientists represent the 'heart of the matter.' They are the people who make the discoveries that the network reports as achievements, who produce the technologies that the network exploits. They are the ones who embrace or resist the organizational goals of the network. Individual scientists – their beliefs, values, opinions – shape the network's research culture. Much depends on the way they locate their loyalties with respect to the various institutions in which they are enmeshed: their hospitals and universities, the network, industry, and 'science' itself. This chapter follows researchers as they undertake, or choose not to undertake, *adventures in the nature of trade.*

This phrase captures an archaic but useful concept that has its technical roots in taxation law. It describes 'in-between' types of activity that generate profit but are not part of a taxpayer's regular business. I might buy a piece of art, for example, or some gemstones. If I am not a dealer in these commodities but I expect to resell them eventually at a profit, tax law says I am undertaking adventures in the nature of trade.

The academic life sciences today seem like just such adventures, somewhere in between public and private enterprise, and somewhere in between the discovery of knowledge and its application. The commercialization of molecular biology remains an activity that academic scientists engage in 'on the side,' as a supplement to their traditional responsibilities.

Then and Now

The genesis of the phrase 'adventures in the nature of trade' lies farther back in the historical record than taxation law, though perhaps a link can be made. The term recalls the sixteenth-century spice trade, when merchant adventurers sought new spices as eagerly as new gene sequences are sought today. Barriers to entry were high. The Portuguese controlled the routes to the Spice Islands in the same way that 'big pharma' controls markets in therapeutics today. England's Company of Merchant Adventurers,[1] founded in 1551, sought an alternative route. Three ships set sail in 1553, but through haste, greed, and ignorance, none reached its goal. Others soon followed, however, despite the risks, because – like 'big hit' biotechnologies today – the potential rewards of the spice trade were enormous.

The Elizabethan equivalent of a blockbuster drug was nutmeg, touted as a cure for the bubonic plague that was decimating London. First-to-market advantages for nutmeg entrepreneurs were huge. Purchased for less than a penny a pound in the Spice Islands, it sold in London at a 60,000 per cent mark-up.[2] A simple sailor could be set up for life, just from what he had smuggled about his person, and a ship's captain would reap untold wealth. Investors earned enormous returns on successful expeditions, but were often wiped out when expeditions failed. Like today's biotechnology sector, this was a high-risk, high-reward, highly speculative milieu.

High rewards, then as now, often depend on protecting the competitive advantage that accrues from intellectual property. The Portuguese controlled the spice trade because they were the only ones who had mapped the route. They owned the navigational knowledge and defended their interests with big cannon. Today's big guns are intellectual property lawyers armed with 'cease and desist' orders.

An academic scientist, however, experiences a high degree of ambivalence when keeping knowledge from others – even when engaged in commercial activities. The need to avoid disclosure until intellectual property rights are secured must be reconciled with a prior commitment to the free flow of ideas. Arie Rip argues that scientists like these, working in commercially sensitive fields, have historical counterparts. During the Renaissance, 'professors of secrets' used their artisanal knowledge to collect and develop 'recipes' that they sold in the marketplace or to sponsors. 'They had to advertise themselves and their knowledge in order to create some visibility. However, at the same time they had to keep their secrets in order to maintain a competitive advantage over other such "professors."'[3]

Today's scientists, socialized in the open science model, are faced with a similar dilemma and respond in two ways: they either embrace change, or they reluctantly accommodate it. The latter attitude, says Rip, is particularly apparent among 'the old élite, the spokespersons for established science.'[4] The stance of reluctant accommodation characterized the attitude of the majority of CGDN scientists to the network's commercialization mandate. One way to begin to understand this attitude is to examine the way network scientists locate themselves in relation to their 'home' institutions.

Localizing Cosmopolitans

The sociological concept of locals and cosmopolitans can help illuminate the 'locatedness' of network researchers. Robert K. Merton applied the term to community leaders, and it was subsequently adapted to describe professionals in organizations and scientists working in industry.[5] The local/cosmopolitan distinction provides a contrast between location in physical space (where you work; whom you work for) and location in cognitive space (professional knowledge, loyalties, and norms of conduct).

Academic scientists are generally thought of as cosmopolitans; they identify more with the institution of science than with the institutions in which they work. Governed by professional norms and peer approval, they embody the ideal of professional autonomy. Indeed, in Glaser's view, cosmopolitan allegiances like these may manifest in 'a concomitant lack of loyalty to and effort towards' their employers.[6] In contrast, scientists who work in industry are 'locals' – salaried professionals, loyal to and constrained by the practical goals of their corporations.

Both are acknowledged to be ideal types, with most scientists representing a blend of the two orientations. The relative emphasis on one or the other depends more on the motivation and direction of their work effort than on any intrinsic qualities of the researcher or the organization. Thus, an industry scientist working on basic research would be cosmopolitan in orientation, whereas an academic scientist working to develop a particular application under industry sponsorship would adopt a local orientation.

This notion of blending seems intuitively correct: every scientist interviewed for this study was a principled 'cosmopolitan,' but all maintained a portfolio of research projects, some of which were more 'local' in orientation than others. To return to Stokes's formulation, outlined in chapter 1, 'cosmopolitans' seem to belong in Bohr's Quadrant; 'locals' are located in Edison's Quadrant; while those with mixed or blended status – 'loyal to science but interested in facilitating the utilization of technical results'[7] – are the Pasteurs of the network's social world. The ability of the Stokes model to accommodate earlier formulations speaks to its robustness. I will return to this later.

These distinctions and overlaps carry a further layer of meaning to do with institutional relationships. Earlier in the study, I made the (understandable) assumption that network scientists 'worked for' (i.e., were employees of) the universities and hospitals that paid their salaries. That being the case, networks were free riding by not compensating the employers for the time researchers spend on network affairs. In the interviews conducted for this study, however, it soon became clear that scientists perceived their status quite differently.

Academic scientists made it clear that they work for no one but themselves. Their response embodied what was earlier referred to as the 'ideology of the autonomous researcher.'[8] Fiercely independent, they regarded their autonomy as a fundamental element of their scientific identity. Indeed, perceived loss of autonomy was one of the major deterrents associated with commercialization. As autonomous researchers, academic scientists considered themselves not employees but 'franchise owners,' funding their research through the grants economy and conducting it under the auspices of the university. Thus, for them, the university was simply 'a place to be a scientist'; a place providing access to laboratory facilities and institutional infrastructure. Not being an employer, the university could exert no moral suasion or domain control over research direction or scientific activities. In other words, it was understood and accepted by the university that cosmopolitans control their own work.[9]

As 'franchise owners' rather than employees, scientists saw no conflict in becoming involved in network activities because *nothing changed* in their relationship to their home institutions. They continued to conduct their research as before, with research council funding and university support. As a senior scientist explained, however, the networks supplied 'far more than infrastructure'; they offered 'an atmosphere, an ability to move into this new world that the university had never provided.'

The local/cosmopolitan distinction remains useful. However, the propensity of some academic bioscientists to undertake adventures in the nature of trade and the blurring of boundaries that occurs when they do so tend to turn the distinction on its head. These people are cosmopolitan in some manifestations and local in others. Clearly, some refinements are called for. A new, three-part typology – one that builds respectfully on the work of Stokes and others – may go some way towards understanding how the organization of research now escapes earlier classificatory structures. In the typology that follows, I will talk about 'merchants,' 'settlers,' and 'translational research' and explain the significance of each term.

Merchants and Settlers

In the days of the spice trade, successful merchant adventurers were masters of context, moving effortlessly between the intrigue of royal courts, the cut and thrust of markets, and the demands of international finance. Taking these individuals as exemplars, I propose the term 'merchant scientists' for the latter-day adventurers who move confidently between the world of academic responsibilities and laboratories on the one hand, and their spin-off companies, contract research, and clinical trials on the other. Merchant scientists are intimately involved at all stages of 'the pipe' – raising venture capital, proving out the concept, scaling up, and promoting the resulting products. Their combination of scientific and cultural capital and entrepreneurial flair allows them to readily accommodate the very different worlds of bench science and commercial enterprise, academy and marketplace. They straddle all boundaries with panache.

While hybrid in inclination, merchant scientists firmly anchor their credibility in the scientific field. Reputational capital – prestige, recognition, and scientific authority – depends crucially on outstanding records of publications, promotions, grants, and awards. And merchant scientists are elite performers by all conventional scientific standards. Analysis of the résumés of CGDN's merchant scientists and the top academic inventors at the University of British Columbia indicate superb grantsmanship, extensive publication records, and peer recognition as intellectual leaders. This reputational capital confers a 'halo effect' on their non-scientific activities.

Others have addressed the phenomenon I call merchant science. In sociological studies of the pioneer molecular biologists who established the biotechnology industry in the United States, Lynne Zucker and Michael Darby found an elite whom they called 'star scientists.' Stars excel in both commercial and academic fields. They are 'extraordinarily creative, innovative, and productive individuals' with the 'vision and genius [to] consciously change the boundaries of what is possible.'[10] In the higher education literature, Sheila Slaughter and Larry Leslie have used the term 'academic capitalism' to describe the policy-driven trend towards commercial activities.[11] Like the merchant scientists in the present study, the 'academic capitalists' they interviewed saw no conflict in having the state subsidize their commercial activities. Rather, because the research was viewed as having social utility, financing it from public resources was uncontroversial. These scientists do not turn away from public goals and towards the market; rather, they elide the two, and define market values as contributing to the advancement of science and the public interest.[12]

But the majority of academic researchers choose not to engage in merchant science. For them, the intellectual costs of engagement far outweigh potential benefits. Unlike the quest for fundamental knowledge, the quest for profit is viewed as tedious and distracting. These researchers make only token forays into the market, just enough to meet the strategic requirements of their grants. This type of lip service to current policy is Rip's 'reluctant accommodation.'[13]

Speaking of basic research ('the republic of science'), Michael Polanyi described it as a society of explorers, a system that cultivates radical progress. At the same time, he said, it is a system deeply rooted in tradition.[14] His insight captures a highly productive tension. Conservative institutional and normative structures underpin intellectual daring, allowing academic scientists a secure haven from which to explore new horizons and exploit new insights. The non-merchant scientists I met at CGDN placed particular value on the intellectual freedom that flowed from their settled status and privileges as university researchers. For this reason, I decided to call this group 'settler scientists.'

The word 'settle' carries a number of complementary meanings. To settle is to become established; to achieve certainty or clarity or confidence; to decide or determine. This 'settler class' of scientists seem confidently established in the pursuit of clear determinations about facts of the world. They are established in the traditional economy of free intellectual inquiry and open exchange of knowledge. At the same time, they are also 'settlers' in the pioneering sense, exploring new horizons and claiming new territories along the frontiers of knowledge. As a group, however, settlers are becoming concerned about changes in their academic havens. While merchant scientists embrace new expectations of commercial and social relevance, settlers fear erosion of unfettered exploration.[15]

What were the relative proportions of merchants and settlers within CGDN? Analysis of network documents and interview transcripts revealed that very few researchers were involved in commercial activities in anything other than a peripheral manner. As a result, they were relatively uninterested in the network's commercial mandate.[16] As the network's CEO said, 'It's a challenge getting them to even *start* thinking about commercial activities or to see translating their science as anything other than a necessary evil. They enjoy the science. And the science allows them to go off on any tangent that they want. The commercial world doesn't.'

All twenty-one founders initially belonged to the settler class. In the intervening decade, three of that group moved into merchant science. But still, only five or six in total of the fifty network researchers were 'first tier' merchant scientists by the CEO's count – those who would be willing to take leave from their academic research to devote themselves full time to commercial activities. About the same number were somewhat interested in the commercial world, he said, but not sufficiently to make even a half-time commitment. At most, then, 10 to 15 per cent of the total complement of investigators were involved in commercial activities, leaving 85 to 90 per cent with 'settler' sensibilities. Since these data are qualitative, not quantitative, there is no question of statistical significance, but the percentage was triangulated against two additional studies – one of university technology managers in four universities, another of Canadian genome scientists – with similar results.[17]

As a heuristic, then, it may be acceptable to assume an 80:20 ratio between settlers and merchant scientists. Rounding up allows a margin for the more junior researchers missed by the study. This is important, because there may be a generational effect at work. Today's senior researchers – merchants and settlers alike – were enculturated in settler labs. As young researchers train under merchants, a shift can be anticipated over time.

The merchant class and the settler class display different dispositions at the margins but share a common goal of advancing knowledge. They converge and diverge around the notions of 'translation' and 'translational research.'[18]

Translation and Translational Research

In the first instance, 'translation' is almost synonymous with boundary work or articulation work. In other words, it has to do with the relation between academic and commercial fields and the way merchants and settlers rhetorically position ('translate') themselves in relation to those fields.[19]

In the second instance – the main focus here – 'translational research' is a set of practices that transports knowledge from the domain of discovery to the domain of use by converting findings from basic and clinical research into tools, resources, or information.[20] It is deployed here as an 'actor's term,' meaning that my informants (scientists and public servants) used it routinely in their discourse; however, I had not seen it in the literature at the time I began to hear it in use.

At that time (2001), a web search revealed only a handful of references to translational research – mostly in policy and program documents relating to cancer research – and few settled definitions. A report from a 1998 meeting of the Breast Cancer Funders Network noted that while many of the participants were interested in finding ways to facilitate translational research, there was no consensus on meaning. A workshop at the Mount Sinai School of Medicine in Toronto ruminated on an interesting distinction between 'transitional' and 'translational' research. (The former converts discoveries in basic science into clinical applications, then uses the resulting clinical observations to generate basic research foci, in an iterative loop. The latter focuses on the integration of activities from bench to bedside.)

The Canadian Institutes of Health Research was just beginning to define itself in 2001 and had adopted 'knowledge translation' (KT) – defined as the exchange, synthesis, and ethically sound application of research results – as a core mandate.[21] KT seemed to share some elements of translational research, but how exactly it would be put into practice was unknown at that point.

Despite these isolated examples, the term 'translational research' seemed to have little general currency in 2001, and I wondered if it was an artifact of local usage. When I repeated the exercise in May 2004, however, while revising this manuscript for publication, Google returned more than 57,000 'hits,' and PubMed was indexing more than 400 peer-reviewed articles in which 'translational research' was a keyword.

The intervention of the U.S. National Institutes of Health was a driver of the heightened interest. NIH wanted public-sector science to start harvesting new therapeutics and diagnostics from fundamental research, an activity that was the traditional province of the private sector. High attrition rates, with attendant costs, had long dogged the translation process. The NIH had been seeking a way to reverse these trends.

In October 2003, after a year of study, the NIH introduced a new 'Roadmap' in which translational research was a key element.[22] The plan would establish a number of new 'Translational Research Centers' to provide sophisticated advice and assistance for scientists bringing a new product from the bench to clinical use. This assistance would include laboratory studies to understand a therapy's mechanisms of action; pre-clinical drug synthesis and toxicity testing; industry-standard manufacturing capacity; and expert advice to ensure that the Food and Drug Administration's drug-development regulations are observed. The Roadmap's facilitative, funding, and other services have attracted considerable interest since its publication.

Complicating any straightforward definition of translational research, however, my empirical data reveal two distinct but related senses. The first is the bench-to-bedside work that translates abstract knowledge into clinical and therapeutic options ('translation to practice'). Some settlers work here. Physician-scientists, for example, through the medium of clinical research, were involved in the interface between research and clinical practice long before commercial considerations intruded. The second sense is the lab-to-market work that translates the same knowledge into commodities and products ('translation to profit'). This is the domain of merchants.

Whether motivated by profit or by practice, researchers who are willing to translate their basic discoveries into useful applications reside, by definition, in what Stokes calls Pasteur's Quadrant. So this domain captures merchants like Hayden as well as settlers who are willing to 'take the next step' by translating their research into practice. However, as will be seen later in this chapter, there comes a point in translation when even researchers themselves refuse to categorize what they are doing as 'real' science. At that point, I suggest, we move into the realm of Edison.

Under market conditions, the two senses of translation blend together. The same process that translates knowledge from the laboratory bench to the patient's bedside also translates it to the market.[23] Whereas translation to practice is an easy 'boundary object' to get

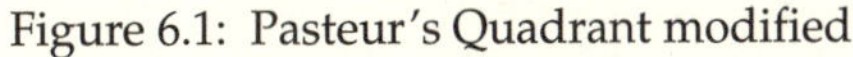
Figure 6.1: Pasteur's Quadrant modified

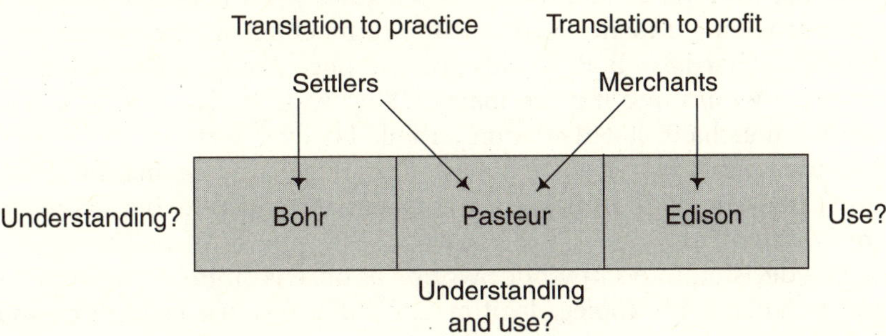

behind, translation to profit was the source of conflicts and resentment within CGDN. It seems, therefore, that it is not the *activity* of translational research that is contested but the *motivation* of the researcher. The slippage between the two meanings requires further study, as does the whole concept of translational research and the space it occupies between discovery and application.

Figure 6.1 illustrates the relationships among merchants, settlers, and translation and maps them against Donald Stokes's concept of Pasteur's Quadrant – the location of research dedicated to *both* understanding and use. As will be seen, 'pure' settlers are located in Bohr's Quadrant. To the extent that they are willing to undertake translation to practice, however, they cross the boundary into Pasteur's Quadrant. Similarly, 'pure' merchants lock into Edison's Quadrant, but their translational work locates them in Pasteur's Quadrant much of the time.

Hearing Voices

Understanding the relations between merchants, settlers, and translation helps to illuminate the ways in which public and private, basic and applied, map onto each other during a period of cultural change. But the tensions and contradictions inherent in these orientations can be profound, as the voices of network scientists will confirm.

Bound in Neither World

When describing the field of merchant science, participants in CGDN used the language of 'worlds.' According to the network's commercial director, merchant science bridges 'the world of the academy and the world of commerce,' and merchant scientists 'walk in both worlds; it's not a problem for them.' As one settler commented of his merchant colleagues, 'They have both worlds going at once. I don't know how they do it ... I would find it more than I would want to do.' And as a successful merchant stated, 'I don't think I'd ever want to be wholly bound in either world, to be honest. I'm quite happy having the ambiguity between both. It makes life complex and intricate. But it's really interesting.'[24]

The decision to occupy both worlds at once is strategic; it provides the scientist with choice, legitimacy, and protective coloration and marries scientific credibility with commercial expertise. Unlike 'real' entrepreneurs in the market economy, merchant scientists undertake no personal financial risk; sinecures drawn from their academic world buffer them from the volatility of the biotech sector. And unlike 'real' medical geneticists in the academy, merchant scientists find that their commercial activities buffer them from departmental responsibilities. Ultimately, a decision for one world or the other may have to be made, but that decision is deferred as long as possible because of the differential benefits that accrue from each. Said one, 'I'm still hiding under the guise of being an academic scientist.' Describing himself as 'something in between,' another said, 'Even though I run a company, I consider myself an academic scientist when I'm in academia. There's no other way to do it.' A third said, 'I prefer to do what I do within the university, for now. But who knows? ... I take it one year at a time.'

At the time of our interview, 'Joseph,'[25] one of three partners in a fast-track U.S. biotech company, was on a leave of absence from the network and his Canadian academic world. He had taken a year out to work exclusively for his firm and pilot it through a merger with a larger European public company. Soon he would have to decide which world to inhabit permanently. The new larger company had offered him the role of leading its U.S. operations. He had not yet decided whether to return to Canada and his academic post. 'We'll have to see. I'm living a week at a time here, I'm not worried.' It would be a difficult decision. He had an international scientific reputation from his academic work, but the commercial milieu was new. 'The bulk of my work, the work I've become known for in the last few years, is completely independent of any commercial activities or considerations.' At the moment, he was running what he described as 'parallel lives':

> One has to do with cancer and bone tumours, and we publish in regular journals like *Nature* and *Proceedings of the National Academy of Science.* That's a very academic pursuit. And then there's the stuff I'm doing with the company, which is completely commercial ... You have to be scientists to move it forward, but I don't consider it to be anything other than a commercial activity. The two lives are completely parallel; they're not connected.

It seems to be the case that Joseph's academic and commercial lives *were* 'parallel and unconnected.' The leave of absence had defused ethical conflicts between his corporate and academic activities. But that principled distancing is unusual.

Ethical Questions

Merchant scientists straddle the boundaries between research and development, science and commerce, the public and private sectors. In a manner that some observers find disturbing, merchant scientists move confidently between their academic and commercial responsibilities. Since these various activities are usually housed in the same premises, potential and perceived conflicts of interest abound.

Most merchant scientists remain in place as faculty members, grant holders, teachers, supervisors of graduate students, laboratory leaders, and clinicians. At the same time, they may also be acting as contract researchers, patent holders, officers of spin-off companies, and directors of clinical trials for 'big pharma.' Not surprisingly, they sometimes lose track of whom they are representing from moment to moment, and that multitasking can generate a number of ethical issues. One informant from a university industry liaison office ruefully referred to his department as 'the office of conflict of interest.'

I visited 'Jim' in his university offices. Several U.S. patent certificates were framed on his walls next to faculty teaching awards and journal covers that featured his breakthrough basic research. It was a scruffy, long-inhabited office that contradicted expectations. Jim, after all, controlled millions of dollars of industry and public funding. He arrived late, sweeping into the room trailed by graduate students and his personal assistant, all of whom were quickly shooed away.

An archetypal merchant scientist, Jim explained that twenty years ago, when the federal government adopted strategic science guidelines, he took them seriously and dedicated himself to creating direct applications from his academic research. Over the years, he had founded dozens of companies to exploit his work, mixing public and private funding and always working from his university base. As director and shareholder of those companies and leader of the academic research projects in which they are invested, he described his life, laughingly, as 'an ethical minefield' and cheerfully admitted to being 'in a conflict of interest on everything I do.'

Jim defended the leveraging of personal advantage from publicly supported research. He argued that public funding is intended as an investment in the knowledge base. In the course of a basic research project, potential products sometimes 'happen' to emerge. When that occurs, 'it's OK to commercialize them because they are inadvertent issues.' In other words, no conflict exists because the funding is not directed to product development and also because profits are relatively modest. University-based start-up companies are 'small potatoes,' so there is no reason to take them out of the lab. Further, the motivation is intellectual and social rather than financial. 'No one is doing this to make money. If you want to make money you have to change your career and move into the company. You commercialize because it's the ultimate extension of your research life – to create something practical that directly impacts people.'

This conviction might be called the 'translation ideal.' It captures both 'translation to practice' and 'translation to profit.' While the ends of merchants may be virtuous, however, many settlers find the means a cause for concern.

> SETTLER: I've seen cases where company people are working side by side with the researchers, where the academic research team are studying the same thing as the company people. And, I think that's pretty tricky and should be [stopped].

> SETTLER: I think a problem can easily develop when you have a company and you're then involved with contracting work back to your own lab. I think it's dangerous.

Joseph, whom we met earlier, described his method of coping with potential conflicts of this type.

> What you try to do is put into place [agreements] that certain things will happen when a conflict arises. In other words, if I discover something on

> the way to work and I happen to be going to the company and my university lab on the same day, the lab always gets credit for it. The university always gets the patent, even if the company is interested in the idea. We never try to split hairs and say, 'Oh, I thought of that when I was at the company.' It just doesn't happen. We always throw the university the intellectual property.

One of the major concerns when worlds overlap is the vulnerability of graduate students. 'Merchant' supervisors can be in a potential conflict if students have been involved in commercially sensitive research that constrains their ability to publish in a timely fashion. This situation was widely recognized as unacceptable, and policies were in place to limit potential delays. Merchant scientists argued that such situations were unlikely to arise and saw concerns as unwarranted.

> MERCHANT: I'm not sitting there trying to exploit my graduate students, getting them to work on company business. It's very different science ... So far [none] of them have the excuse that they can't publish because the company's keeping a lid on it. In fact, they had better look to themselves as the reason why it's not published, not the company.

But some graduate students *do* conduct their thesis research on company-funded projects. And sometimes the company's economic or commercial goals change, and projects get redirected or cancelled.

> MERCHANT: You end up shifting resources away from one thing to another in a lab. And a graduate student's doing his work, and he finds that the resources around him are diminished because it's gone to another project that has major commercial value. That's where there's conflict, resentment, bitterness. Those things happen.

Even postdoctoral fellows can be vulnerable if they don't know, in advance, what they're getting into. And they don't always know.

> SETTLER: Sometimes they come as an academic postdoctoral fellow to work on a project for Dr So and So. And when they're six months into the project they realize that 'Hey, this project really is one of the prime goals of Dr So-and-So's company, and he is going to make a bundle of money from this if I figure it out.'

As well as experiencing conflicts of interest, merchant scientists also tend to get caught in conflicts of commitment. Within the academy, a scientific discipline operates as a 'workplace culture,' a type of moral economy in which 'tacit rules of mutual obligation guide community life.'[26] But the commercial activities of merchant scientists override these tacit rules; departmental responsibilities assume a lower priority, attracting the resentment of settler peers.

Joseph found peer resentment a major problem. 'The people in your department often don't fully understand it and are generally pissed off. It's serious and never-ending.' Part of the antipathy stemmed from the fact that he was seen to be working in his own company for his own benefit. It was he, rather than his academic community, that would realize the rewards. Another aspect was that Joseph was rarely available to work in the department. 'Someone else has to teach for you; someone else has to be on committees.' Joseph perceived the disparity in compensation as an issue also, given the relatively low pay in Canadian academia. 'People in companies get paid four to five times the salary they'd be making at the university.' Add these facts together and 'they build some pretty serious resentment.'

Conflicts of commitment were especially apparent when merchants partnered with powerful pharmaceutical companies. Michael Hayden, for example, had a long-standing relationship with big pharma. Merck underwrote a significant portion of the capital cost of the Centre for Molecular Medicine and Therapeutics (CMMT). The company also funded the clinical trials he ran and sponsored a number of his research projects. As one federal observer explained,

> Merck always wanted Mike Hayden. They couldn't recruit him, but they bought him in other ways. What Merck got eventually was more than Mike Hayden; they got the network. Or a good part of the network. But that's really the whole point of the program. And the partnership has to be mutually beneficial.

As Hayden acknowledged, 'When Merck calls me to do something and I've got a graduate student to meet, what am I going to do?' Merck calls on his influence whenever a pharmaceutical industry issue is on the table. Although he prefers not to get involved, he will usually do what they want, if he agrees with them on the issue. The truth is, he said, 'I want them to be happy. My future and our [the research team's] future depend on them.' Hayden sees Merck as worthy partners. 'They've got a lot to add. So it doesn't help us to not fulfil some of their requests, if they are science driven.'

That phrase – 'science driven' – is the key to resolving conflicts, according to Hayden. The agreement with Merck, for example, gave him total discretion over the direction of research. 'I choose what we work on. In the contract, it doesn't say what we're gonna do. It says that Dr Hayden, in consultation with his colleagues, shall decide what research will be undertaken.' Conflicts occur, he said, when relationships are 'dollar driven,' because corporate cash can weaken research ethics. When relationships are science driven, and supported by strong and clear agreements, conflicts are under control. But Hayden acknowledged that much depends on the strength and integrity of the individual and his or her ability to withstand temptation.

> If you're not able to compete, and not able to get sufficient funding for research, you end up taking money from industry because you're desperate. And you end up taking any money from industry ... because you're helpless and weak and you haven't been able to raise money elsewhere. And so you end up doing things that you really shouldn't do.

To minimize both the appearance and the potential for conflicts, the 'right' thing to do is move a start-up company out of the lab as soon as possible and for the scientist involved to take a leave of absence from the university, as Joseph did. This puts some physical distance between the merchant scientist's 'public' and 'private' worlds. Most universities have set up research parks and incubator facilities for this purpose. But that still doesn't solve the problem. The basic issue, as one distinguished scientist-observer comments, goes much deeper.

> It is a contradiction in terms for a scientist to be an entrepreneur. One is trying to make an industrial product, a market product, and the other is trying to do good science. It reminds me of the saying – you can't serve God and Mammon. There may be some instances where the two are congruent and you're lucky. Particularly in the non-health areas, where people can decide for themselves whether a new and different gidget is in everybody's best interest. But in health product areas, it's much more difficult. Because your motivation is to push that product, and the public can't assess whether or not it is useful and good. There's a knowledge

gap. And so the chance for the public interest not being served by the union and marriage of scientific pursuit and entrepreneurial pursuit is greater – there's a problem.

Incorporating Merchant Science

By the end of my fieldwork in June 2001, CGDN's merchant scientists were actively involved in nine spin-off companies, six of which were launched in Phase III. Of the three prior companies, Apoptogen,[27] founded in 1995 to exploit the discovery of apoptosis inhibitors, merged with another company to form Aegera Therapeutics Inc. and a subsidiary, Aegera Oncology Inc. As yet, the company had no products in clinical trials, but current capitalization was approximately $28 million. Neurovir Inc., launched in 1996,[28] was exploiting herpes as a viral vector as well as pursuing the potential of a 'basket' of technologies bought in as a portfolio. The company had a product at phase two clinical trials and had recently merged with a larger European company. According to a corporate informant, the company was then worth 'slightly more than $100 million U.S.'

Xenon Genetics Inc.,[29] established in 1997, exploits genes associated with lipid disorders. In May 2000, the company signed a deal potentially worth $87 million with pharmaceutical giant Werner Lambert. Originally capitalized at $13.2 million, a 'mezzanine financing' deal valued at U.S.$47.6 million was completed in May 2001. In November 2000, Xenon took over RGS Genomics Inc., a Phase III spin-off founded by three network researchers from McGill.

Newer companies began to establish themselves. Solutions By Sequence Inc., was set up in 2000 to undertake retinoblastoma testing.[30] The partners had modest aspirations in terms of financing and profit; the primary goal was to service the at-risk population. Genexyn Pharmaceuticals Inc.[31] (2000) pursued gene and protein trapping. Signalgene[32] (1999) was capitalized at around $50 million and had negotiated a $1.2 million research contract to extend work on osteoporosis and psoriasis. Ellipsis[33] (1999) had ambitions as a gene identification company, and EcoGenix Inc.[34] had raised its first $750,000. Certainly, since Phase III and the appointment of the new CEO, increasing emphasis was being placed on company creation.

Perhaps relatedly, by the summer of 2001 CGDN's board had become increasingly concerned about perceptions of conflict of interest, as had the NCE directorate. The essence of the problem was the start-up companies. As products of network activity, these companies tended to become industrial partners. The founder scientists became involved in the company's first-generation product development, using venture capital raised for that purpose. At the same time, however, they remained as network scientists, and the network continued to fund the academic element of their work. The opportunity for real conflict of interest could arise if the company or its scientists used their relationship to leverage money out of the network.[35] Full disclosure of interest and arm's-length relationships were essential, but these matters had always been handled on an ad hoc basis.

Now, however, the NCE's conflict-of-interest framework had been adopted as part of the requirement for the Phase III network agreement, signed in May 1998. That framework charged the board of directors with the responsibility of 'managing conflict of interest, and determining and implementing the appropriate course of action.'[36] With the mid-term review pending in May 2001, the board was warned by the network's new commercial director[37] to expect questions from the review panel about how the conflict-of-interest framework was being implemented. They were advised that unless a process was put into place, the review panel might identify a deficiency. Accordingly, just before the review, the board appointed a pro bono conflict-of-interest officer (a lawyer and former hospital board chair) to advise it on potential conflicts. The board also directed staff to prepare a list of the direct and indirect financial interests and positions of influence of each individual in the network, including scientists, board members, and professional staff.

'Excursions into the Land of Ignorance'

Settler scientists are the beneficiaries of publicly funded and curiosity-based career research programs. Rather than venturing out into the commercial world in pursuit of profit, settler scientists undertake adventures in place. They mine knowledge, making what 'James,' a distinguished biochemist, calls 'excursions into the land of ignorance.'

> I am a career academic scientist in the medical field. I've always worked out of the university and its frame of reference and its culture. And I've always been funded by [the state] ... That's the only thing I know. I was allowed to be a scientist. I was allowed to make excursions into the land of ignorance, to try to bring back knowledge that would benefit patients.

James and his colleagues developed many technologies but patented none of them. Patenting was simply not an option. 'We just never did that then. We just got on with the job of science.' Science was a privilege and its own reward. He was allowed to do research, he enjoyed doing research, he was paid to do research and could continue to do it if he satisfied his 'paymasters': peer reviewers and the deans of the department. He contrasted that environment with the market-driven pressures of today. 'The paymasters now are the shareholders. And the venture capitalists. And that's why I have had trouble with the network in its commercial direction. The emphasis on profit as what we have to do makes me feel uncomfortable.'

James's resistance to commercialization is principled, but others are more focused on practical considerations. Why would they compromise satisfying academic careers to get involved with industry? Why would they risk venturing into an area where they have few skills?

> People are worried that it's not going to work out if they give up their academic career and go into industry. They are leery; they don't feel all that confident. After years and years of writing grants you get to know the system. To now go out into the business world, it's a very strong break to make.

Many settlers view commercial activities as a subsidiary requirement of network membership, a chore for which they receive a grade on their annual 'report cards': 'My report card was always stars for science and good for networking and zero for commercialization,' one researcher noted. They may feel guilty about not doing more commercial work, but their own research programs come first – 'I really have not had any time left over to pursue commercial interests. It's something I would do if time allowed.' Settlers who do participate on the commercial periphery – for example, by licensing their discoveries to others – do so to underwrite their discovery-based research programs. 'We're not doing this [patenting] out of personal gain. I don't think any of us have any illusions on that score. With monies being so tight right now, if I do make anything, it would just go rolling back into the lab.'

Some are peripherally interested in what merchants do but are deterred by the time commitment required. The participation of the researcher is essential in any move to capture the commercial value of a discovery. Even in relatively simple arrangements, such as filing for patents and licensing them out, the researcher must be involved. And that involvement takes away time from the laboratory.

> Sometimes, when they take the discovery to the lawyer, I almost dread them coming back and saying, 'Oh, we should patent this.' Because I realize how much time that's going to take. If they are good lawyers, they can come up with most of the writing and stuff like that. But ... it still comes back to you doing the detailed work.

Another source of the settlers' resistance to commercialization is the loss of freedom that accompanies ventures outside the academy. Academic researchers are highly self-directed, and the academic environment allows them almost complete domain control over their scientific activities. As mentioned previously, in true cosmopolitan fashion one described himself as a 'franchise owner' rather than an employee. He chose to locate his franchise (grants and contracts) at the university, but the institution simply provided 'a place to be a scientist.' He could move elsewhere at will. 'Franchise arrangements' were similar in all locations, providing access to university facilities, salaries, and infrastructure, in return for teaching and supervision of students and other types of service.

This degree of independence fosters curiosity-driven research and the freedom to follow serendipitous directions. It is one of the 'fun factors' in doing science: 'discovering stuff ... and saying, "Hey! No one's ever seen this before."'

> I can wake up tomorrow morning and say, 'Wow, I'm going to try to get a grant on juvenile diabetes. I've got a crazy idea but it might work and if I get the funding, we're actually going to try to do it.' And that freedom is just the greatest thing.

For settlers used to this level of autonomy, the constraints of commercialization are unacceptable. The accountability of merchants to their 'stakeholders' – venture capitalists, 'angel' investors, biotech companies, big pharma – is viewed with horror. Merchants have to tolerate close scrutiny from these interest groups. Seemingly arbitrary 'milestones' have to be achieved to release more funding. There is little in this prospect to attract a grant-funded settler.

> To have a bunch of investors tell you, 'Well, you haven't met your milestone,' is not quite right. If you've got an academic lab and you're also in

> some company and the board of directors is saying, 'Well, you said you were going to have this done by now, what's going on here?' And then you've got to go and yell at your scientists to get them going. Well, you know, that kind of pressure, to me, is not worth it. Some people like doing all that stuff. But I would find it just too stressful.

Some settlers initially embraced the new commercial relationships but found the demands of investors unacceptable and retreated to the academic environment. One, for example, actually made the move and spent several months setting up a business enterprise to commercialize his research. But he soon became disenchanted. 'I didn't like what was happening, I didn't like how it was happening. [And] the funding that I had expected to get from the venture capital group didn't come through.' He realized that this was not what he wanted to do and jumped at the offer of a new academic position at a different university. 'They wanted me to develop a new research group there, and they offered me $23 million to set up the new facility, so I went.'

Over time, settlers came to accept the inevitability of commercial outcomes. 'We're not trained for it, and many of us view it with a lot of concern because we know some of the horror stories. But on the other hand, we know that we're probably going to have to go that route.' At the same time, settlers have also come to accept the restrictions on disclosure and dissemination that accompany intellectual property protection: 'I think all of us are now more aware not to disclose prior to patenting,' said one. For another, 'Remembering to protect the intellectual property before we publish, that's not so much a strain now as it was ten years ago. We're getting used to that.' Settlers will even move towards translational research themselves, taking out patents and participating in simple licensing deals, as long as they can turn their discovery over to someone else to develop. As one said, 'In the old days, if you made a discovery that might potentially have commercial applications, it would have ended there. Now, you have the possibility of moving forward with it, and the network is there to help us through that.'

Although willing to accommodate intellectual property requirements, many settlers find that the secrecy demanded by the network's commercial activities remains a source of disquiet. They have to be constantly on guard against accidental disclosure. In a small field, simply mentioning the name of a principal investigator can be tantamount to telling a competitor what has been discovered, because everyone knows what everyone else is working on. That competitive aspect undermines the culture of open sharing.

> It was much more fun before. Now [when you go to network meetings] there are people saying 'I cannot talk about this.' Or busy answering their cell phones. So instead of being a dedicated time to talk about science, people are side-tracked. That is the thing that I find different, that I regret. I regret the drive towards commercialization.

Commercial constraints are antithetical to the open science model under which settlers were socialized. Academic researchers have not learned how to extract information from each other with skill under these circumstances. But as another settler pointed out, the constraints can be justified. 'I don't think it would be fair for a company to take all the work ... that we put into a discovery and then make money on it. Without patenting and licensing, anybody and everybody could use it and earn money commercially on our efforts. That doesn't seem fair.'

The irony of this situation is that, until recently, the whole point of public funding for academic science was that 'anybody and everybody' could then use it. Under the postwar social contract for science, government support helped to generate knowledge that was then made freely available to the manufacturing sector. Under the linear model, the resulting innovations would fuel growth and jobs and return taxes to the state, which would fuel the pipe again. Until quite recently, therefore, scientists' rewards came in the form of research support and peer recognition rather than as profits from proprietary knowledge. As shown in chapter 1, 'open science' economists argue that in making knowledge proprietary at such an early stage of the pipe, universities and public-sector researchers choke innovation and stifle wealth creation.

While that question remains open, settler scientists accept that the moral economy of science is changing, and they have learned to work with the new norms. The researcher quoted above on the topic of fairness admitted that, although she was far from enthusiastic about the concept of intellectual property, she could not see a way to avoid patenting her discoveries. 'When everyone else is patenting and licensing their discoveries, one cannot say, Oh well, everybody else in the world is doing it, but I'm not going to.'

So as figure 6.1 suggests, open science and proprietary science occupy a continuum: settler scientists at one end, merchant scientists at the other, creeping towards each other. Justifying this 'creeping commitment' is a rhetoric of translation: translation from the laboratory to the market creating wealth, and from the bench to the bedside supporting health. In justificatory discourse, the two are often conflated. The argument is that translation to practice requires translation to profit: that the production of beneficial technologies requires the motivation of profits. Whether or not that is the case, the desire to translate results adds a new dimension of utility to research programs, achieving one of the goals of the NCE program and situating translational work clearly within Pasteur's Quadrant. However, the scientists themselves – merchants and settlers alike – clearly differentiate between research and translation, discovery and development, using classic boundary-work strategies to make a sharp demarcation.

'I Wouldn't Call It Science'

Translational research is the ampersand in R&D. It involves pre-clinical and clinical studies to test new drugs and procedures (see figure 5.1, p. 140). Basic researchers, and even some merchants, take pains to differentiate it from their own activities:

> MERCHANT: The stuff that we do in a commercial setting is really technology development. I wouldn't call it science.
>
> SETTLER: If you're not doing [basic] research, you're like a mechanic that repairs cars. You may just work on Mercedes-Benzes and it's highly specialized work, but you're still just repairing cars.

In other words, basic science is 'real' science; translational science is not. All scientists interviewed made this distinction, even those who were themselves involved in translational work. As one explained, 'After they have discovered the gene, and they have figured out what the protein is, translating it is really boring; it's not exciting, like basic science.' Another described his company's translational work in the same deprecating tone:

> We develop leads, and do structure/activity relationships, and do chemistry, and build new compounds, and then test them back and see which one gets better and better. That's going to be an exciting drug, but as basic science it's not all that interesting. It's not what's published in the basic

> journals. It's not how they find out how this gene works, and what pathway, and what it interacts with, and so on.

The valuing of basic research above translation and application is deeply embedded in scientific culture. As discussed in chapter 2, one of the goals of the NCE program was to change these norms and, to a large degree, that was achieved. Within the network, much of the stigma was removed from translational research, largely by making it legitimate for scientists to maintain multiple roles. In other words, researchers came to understand that they could do basic research *and* clinical research *and* translational research; it was not an either/or proposition, and all activities could benefit from the interaction.

The program goal of promoting translational research was not understood at first, especially in the first phase. At that time, researchers viewed the NCE program simply as another funding source for basic research. According to Michael Hayden,

> It took a while for everybody to understand the *place* of the NCEs. The place of the NCEs is *not* fundamental research necessarily. There may be some, but it's really the *translation* of research into products and services for economic benefit for Canada. But that wasn't appreciated early on. Nobody really understood. And nobody trusted it.

In fact, many researchers actively resisted the idea that their work would be applied. Gradually, however, through a combination of resource steering and peer example, attitudes began to change. Over time, researchers came to accept that it was a valuable and even scientifically interesting enterprise.

A core of network scientists had always been involved in activities akin to translation; for example, about one-third of network investigators are physician-scientists. The rigours of maintaining a clinical practice as well as a research laboratory may well require people who are more entrepreneurial or opportunistic than the average scientist. But, that speculation aside, studying diseases in situ, and creating therapies, was an honourable goal long before profit became a consideration. A venerable tradition marks supply and service through various small enterprises. For example,

> SETTLER: I sort of run my own diagnostic company on the quiet. It is not a formal company ... The money comes back into a fund within the hospi-

tal. After I have paid for the technical help and the tissue culture, I use the excess to do research. This is all service oriented. I don't do this to make money. I do it because there is a demand for it. If I didn't do it there would be a big hole, which the commercial companies would not fill. They can't make money at it. It is an orphan service.

SETTLER: About twenty years ago ... it was absolutely impossible to buy any of that enzyme ... [needed to treat an 'orphan' disease] because nobody could afford it. So what did we do? ... In a little room on a back street in Montreal, we set up the potential to produce a world supply of this enzyme.

But outside these traditional enterprises, translation was little respected and had always been difficult to fund. It was not what the research councils understood. That was why network support was so important. As one merchant commented, 'MRC thought translational science was pretty boring stuff. And all the agencies thought that was pretty dull. And so it might never have happened without the network push.' Until quite recently, universities and hospitals, as well, provided little support for translational research, and in some institutions that is still the case. Outside the boundaries of the network, translational researchers must still compete with basic scientists for resources and respect. One researcher explained that, in her hospital, basic scientists and administrators openly questioned the propriety of translational work.

[They] would say, 'Well, what are you doing that for? We shouldn't be doing that in this hospital.' That is what they still say at my hospital. 'We should only be doing basic science. Shouldn't drug companies do that [testing]? Shouldn't industry do that?'

Michael Hayden disclosed that it was in part his attraction to clinical work that motivated him to found the network. His research interests and career choices had always closely connected with patient care and the improvement of patient well-being. 'I was pretty much based in the area of translational research, right from the beginning.' As another senior researcher with a similar clinical background observed, 'Translational science has always been my focus. My interpretation of the network was that it should actually make health better.' For physician-researchers, the logical extension of their work is creating a product or service that will have an impact on human health.

> We want answers. What we're trying to answer right now ... is whether inhibition of a certain pathway has prospects for therapy. Well, believe me, the quicker I get that answer, the more I'll feel I'm fulfilling my responsibility. And if we can do it one day earlier, we should do it one day earlier. Industry loves that; they say that's great.

Clearly, in the knowledge-based economy, translating research into products that may improve human health also translates into the potential for significant personal and corporate wealth. Hayden admitted that the network was 'becoming more entrepreneurial, more capitalist.' Thus, it becomes increasingly difficult to tease apart these two types of translation. The profit potential in bioscience 'hits' (in which big pharma and/or the stock market invest in a discovery) is huge. Researchers who pursue translation can become seduced by the economic value of their work. In terms of recognizing the normative changes underway, Shapin suggests that there is a need 'to produce a post-Mertonian picture of the moral economies of science.'[38]

Summary: Market Values and the Life Sciences

This chapter locates the role of merchant scientists in relation to that of settler scientists, points to the key role of 'translational research' in differentiating the roles, and analyses the different values attached to each of these categories.

Philosophers have long distinguished between different types of values.[39] C.I. Lewis identified five kinds: utilitarian, instrumental, inherent, intrinsic, and contributory.[40] A more basic distinction combines the first two types as *instrumental values*, encompassing use for a purpose and as a means to an end. The latter three are classified as *intrinsic values*, which include aesthetics, things that are good in their own right, and things that are good because they are parts of a whole. Briefly stated, 'Things which have instrumental value are good because they can be used to obtain something else. Things which have intrinsic value are good for their own sake, and as intrinsically valuable, they are not exchangeable for something else.'[41]

In a globalized economy, a third type, *market values*, can usefully be added to this typology. Market values are tied to the emergence of classical economics, which defined them as natural and providential and justified faith in the 'invisible hand' – letting the market decide. Market values determine the price to be attached to *instrumental* values. *Intrin-*

sic values resist pricing, however, since partitioning an indivisible whole into priceable components is almost a contradiction in terms. But it *can* be done, and once something of intrinsic value *is* partitioned, the market can assign instrumental worth to it. Bowe says that 'the more we commit ourselves to an ideology that sees value only in instrumental parts, the further we lose sight of the intrinsic value of wholes.'[42]

These concepts provide a useful heuristic for thinking about the conduct of science. Basic science has intrinsic value, and applied science is instrumental, whereas the life sciences are increasingly characterized by market values. In the typology of network scientists that has been developed in this chapter, pure 'settlers' seem to be guided by the intrinsic value of science *for its own sake*, 'settler-translators' appear to understand the instrumental value of their research in terms of human health, while 'merchant-translators' see the instrumental value of health in economic terms. Pure 'merchants' seem to be driven by the market value, or price, of what is translated.

The goal of strategic science policy has been to 'manufacture' merchant scientists by rewarding instrumental research that is geared towards market values – sometimes at the expense of intrinsically valuable basic science. In recent years, basic science funding has been restored, but grants still carry a 'market rider' promoting intellectual property protection and commercial exploitation. Even our universities have been colonized by market values, to the dismay of some and the approval of others.

The climate created by these activities may convey a sense that commerce is taking over academic science. But the results reported here suggest that something different is happening. After a decade of incentives, only a minority of CGDN scientists were of the merchant class. The 80:20 ratio of settlers to merchants appears to resist the state's best attempts at recalibration; funding-related cultural shifts seem superficial at best.

These findings suggest that intrinsic values are not 'up for grabs.' As such, the policy direction may be fundamentally misguided. Market values favour short-term instrumentality over the long-term accumulation of enduring, public knowledge. But *both* are required. Instrumental knowledge usually emerges from fundamental discovery; esoteric, apparently use*less* research contributes, over time, to use*ful* products for combating human disease.

Furthermore, not all instrumental solutions to human disease come in the form of market-friendly 'products.' Research on the socio-economic and behavioural determinants of health show that non-pharmaceutical interventions – such as lifestyle changes, the amelioration of poverty, or early childhood education – have profound effects. But these interventions are not commodities. They are 'public good' solutions in the social – rather than economic – sense, and research on them attracts little interest in the present policy climate.

> Funding is badly needed for research on new and effective ways to get people to change behaviour, and for research on policies that provide incentives and support for healthier behaviours at a population level. We need much more research in these areas, and there is relatively little funding for it.[43]

In light of these factors, the policy focus on 'manufacturing merchants' seems short-sighted at best, and the lost opportunity costs may be significant. Costs, conflicts, and unintended consequences for the research culture should be taken into account when assessing the overall impact of strategic science policy.

7 NCEs and the Public Interest

The case study of CGDN and the NCE program generated a set of six broad themes I consider to be of theoretical, empirical, or policy significance. Another researcher might well interpret the data in a different way and take away a different set of conclusions. Nevertheless, I believe the evidence presented in the study supports the robustness of these findings.

Case Study: Conclusions and Implications

Theorizing Translation

The concept of *translational research* may be theoretically significant. The results of this study suggest that neither 'basic' nor 'applied' accurately captures the empirical reality of much of the work in biomedical research. Nor do these terms apply to the long-standing 'third space' between the bench and the bedside, the laboratory and the clinic, where this work occurs. Translational research – whether 'for profit' or 'for practice' – seems to fit within what Donald E. Stokes describes as Pasteur's Quadrant, where research is dedicated to both understanding and use.

Linking the various explanations of the third space are two common elements: the goal of advancing the public interest or common good, and the blurring of public and private interests in pursuit of that goal. In other words, research in the third space in some way advances the aims of society and the state. It is fed by both science and technology, not by a linear flow that constitutes it as a way-station between basic research and application. In the Cold War years, for example, this sector provided a home for 'mission-oriented' defence research in the United States, largely conducted by private-sector contractors. Agricultural research into new crop strains belongs here, as does the search for new therapies in the biomedical sciences. In many senses, Vannevar Bush's 'linear' model wrote this third space out of history, but it never really went away. A major finding of the present study is the importance of this productive zone at the intersection of former divides. Translation is not a linear process.

Although translational research is now an accepted part of clinical practice and drives many funding decisions in the health sciences, the role of translational research, in the sense used here, does not appear to have been explored in the science studies or science policy literatures. Many physician-scientists in the present study treat translational research as a 'boundary object'[1] or 'articulation work,'[2] differentiating it from 'real' research. However, the NCE program and newer initiatives are making translation an increasingly 'respectable' activity for scientists to undertake.

The results of this study show two distinct but related meanings: translation to practice and translation to profit. 'Settlers' are more comfortable with the first; 'merchants' are driven by the second. Whereas for most scientists translation to practice is a relatively painless transition, translation to profit is a common source of conflict and resentment. It seems, therefore, that the problem is not the activity of translation but the motivation of the researcher. The boundary between the two meanings requires further study, as does the whole concept of translational research and the space it occupies.

As well, in practical terms, translational research seems to 'fall down the cracks' between health policy and science policy. 'Translation to practice' is clearly a significant cultural component of medical research, yet it is not accounted for statistically.[3] For policy makers, an activity that is not counted does not count. With the health sciences becoming increasingly significant in economic terms, however, it is important to measure and understand the activity of translational research and its extent, as well as the market mechanisms that deliver translated therapies to the bedside.

Spatial Dynamics

Throughout this book, examples have been provided of how thoroughly CGDN's configuration was permeated with power relations and exclusionary criteria. These issues were explained as relating to the network's spatial and structural dynamics. The first theme related to the 'clustering' of regional distribution in three main nodes, caused by the 'spoke and hub' configuration. The second related to issues of elitism and equity and the way exclusionary criteria were used to homogenize network membership. The third related to the worrying absence of social reflexivity and lay representation in a network dealing with the social and ethical risks of medical genetics. A fourth related to the network's cavalier attitude towards fiscal accountability. The full arguments, articulated in this book's preceding chapters, carry a number of policy implications for the structuring of future programs. Ways must be found to maintain variety and heterogeneity. A maximum diversity of participants and input is necessary for a healthy research system.

An initial spatial-structural problem with the analysis concerned terminology: because NCEs were *called* networks, it seemed reasonable to expect the weblike patterns and dynamic spontaneity described in network theories.[4] But CGDN resisted this characterization. Many attributes of a network identified in the literature *were* present, as is clear from the analysis. Yet, in many ways, 'the network' did not behave the way networks are supposed to behave and did not look the way networks are supposed to look (the 'spoke and hub' pattern is one example).

Two reasons can be suggested for this recalcitrance. First, the label 'network' is an appropriation, not a description: NCEs were described as networks before the fact. The federal government wanted networks; so, by an act of naming, networks were mandated into existence. But they were empty spaces awaiting time. 'True networks,' however, are historical achievements. They are identified after the fact. People map past connections, record relations, and describe what they find as 'a network.'

Second, theoretical networks are dynamic, open structures that spread across space and time. Permeability is a defining element; networks are unbounded. But because NCEs did not develop 'naturally,' their morphology is different. Unlike the heterarchical, transdisciplinary networks described in Mode 2 formulations, for example, CGDN has more in common with the Mode 1 'command and control' monodisciplinary model of academic science. Also, although it is open in its internal dealings, CGDN is a closed system. Admission is 'by invitation only'; thus, the network is bounded. Later networks, however, particularly those funded from Phase III onward, do seem to have more of the attributes of 'true networks': they are more interdisciplinary and work more in the context of application.

A key element in the conceptualization of networks may be the factor of *incorporation.* Before incorporation, the NCE networks were genuinely 'ephemeral organizations,' acting simply as facilitating agencies for their members; they were not 'a body' (corporeal or corporate). When they took on the corporate form, they also took on the desire to have their own assets, funds, and future. It was no longer enough to be an 'enabling technology' for others. Philosophically, a corporate structure made a huge difference – one that Industry Canada had wanted from the start. As ephemeral organizations, the networks had been dependent on the university. In a way, they could be considered an extreme form of traditional 'sheltered' faculty enterprise. But once they were incorporated, they became separately institutionalized and took on an entirely separate trajectory. They were no longer the sum of their parts.

Finally, much of the boundary work and territorial politics between NCEs and host institutions concerns a determination of who is the more competent at research management. But it is 'a narcissism of minor differences,' because the networks could not exist unless they were sustained by universities, hospitals, and research council funding.

Cultural Norms

Another analytical problem highlighted by this study is the abundance of cultural contradictions in the world of scientific research. These contradictions relate to the tensions between scientific and bureaucratic rationality, collegiality, and commerce that lie at the heart of the NCE 'system.' In a sense, networks are of two minds because of their dual commitment to scientific excellence and managed research. The oscillation between scientific and bureaucratic rationalities shows clearly in the network's funding proposals, which are carefully crafted both to satisfy traditional disciplinary criteria and to demonstrate performance against secular standards such as managerial quality and commercial relevance.

Robert K. Merton's traditional scientific norms – communality, universalism, disinterestedness, and organized scepticism – are honoured *within* the network, where high-trust, high-familiarity climates promote openness and sharing. But they are discouraged by administrators in *external* relations, where secrecy and distrust more often prevail. Knowledge is treated as a communal resource within the network but is proprietary outside, 'protected' by administrators as intellectual property. Merchant scientists embody the tensions, expressing allegiance to traditional norms of academic science while striving for the goals encouraged by the counter-norms of commercial science. Translational research is itself a counter-norm, since it serves the interests of commerce and therapeutic communities rather than being 'disinterested' in the Mertonian sense. Translational researchers (settlers as well as merchants) profit financially from their work, as does the network itself.

In the CGDN, the open, public academic values of its scientists were in conflict with the closed, private corporate norms of its professional staff, who reflected the bureaucratic rationality of the program's gatekeepers. A culture of administrative secrecy prevailed over a political rhetoric of openness and transparency in government and cultural norms of openness and transparency in science. Network scientists encouraged the case study project; they were eager to discuss their work and to share information with a fellow researcher. On the other hand, administrators in both Ottawa and the network did everything possible to discourage enquiries. Requests for information were refused in the name of corporate confidentiality, even though the private-sector partners made no such demands and maintained a fair amount of distance from the network's day-to-day affairs.

Overall, the key policy issue for the NCEs is accountability. The administrators' lack of openness and full disclosure results in misinformation and impedes the adequate evaluation of the effects of policy. At both the network and program levels, obfuscation, obstruction, and the manipulation of information restrict broad and informed discussion.

Network Effects

In assessing the effects of the NCE program, we lack a 'compelling counterfactual'[5] – we do not know what would have happened if no NCE program had existed. Some $760 million in public funds were recorded as invested in the NCE program between 1991 and 2001; insiders suggest that three times that amount, or more than $2.25 billion, is closer to the actual figure. Until recently, however, the NCE program's reporting requirements were so casually policed that, despite this substantial public investment, it is impossible to determine with any degree of confidence what the program's incremental returns actually are. There is no way to calculate which of the claimed benefits are attributable to the program and which would have occurred in any event, given the context and conditions. There is no way to gauge the opportunity costs of allocating more than $2 billion in public funding to one area rather than another. Since we do not know what benefits would have been generated in the absence of the NCE program, we have no way to calculate the value added by the program.

For example, no one knows whether the network system has produced more discoveries, patents, and publications than would otherwise have been the case. ('Probably not,' admitted one of CGDN's founding scientists.) However, this study's data suggest that the quality of research may be higher within the networks, and that the *culture of collaboration* is a key legacy. It is clear that CGDN has had a salutary effect on the way medical genetics is conducted in Canada. Further, in CGDN at least, the material and intellectual resources marshalled by the core facilities constitute much of the 'added value' of the NCE program.

'Public,' 'Private,' and 'Profit'

In the case of the NCE program, determining the legitimacy of public funding allocations is problematic because of the difficulty of calculating the rate of return on public investment. The program has historically emphasized the need to commercialize results and, to the extent that these efforts are subsidized, the state is intervening in the market. Network scientists are financially supported to start up private companies that will act as receptors for discoveries that were themselves publicly funded. Further, the private start-ups are generally incubated within public universities. Thus, new, high-risk private companies are triply subsidized with public money.

The justifications for this funding mix are not clear. In the first case, state interventions in basic research funding are normally justified on economic grounds of market failure (i.e., private companies find it impossible to recoup adequate returns on investments in basic research because the nature of the product invites free riders). The difficulty of capturing benefits relates to the fact that basic research is *non-rivalrous* (once produced, everyone benefits) and *non-excludable* (once created, everyone has access to it). Further, the social rate of return is higher than the private rate of return, so private actors have little incentive to invest. In these circumstances, without a mechanism for collective action, the market underinvests in basic research, and the knowledge base erodes. Thus, the state steps in to fund basic research as a public good, in order to recoup the social returns. To the extent that NCEs generate basic research and publish findings in journals and at conferences, they are producing public goods for the private sector to develop. This is the 'open science' model in action.

It is a different question whether the NCE program is a better allocation mechanism for basic research funding than existing programs, such as the granting councils. The question to be considered here concerns the commercial activities of NCEs and the utility of funding private goods through public mechanisms. If goods are rivalrous and excludable (i.e., the benefits can be captured), the arguments for state intervention dissipate and the market becomes a better mechanism of allocation. That being the case, why are 'public' networks being funded to generate 'private' goods? Does the private sector view this subsidy as unfair competition?

In a sense, NCEs (and universities) internalize the private sector by protecting their own IP and creating companies to exploit it. Almost invariably, these start-up companies are sheltered from market forces in incubation facilities within the university. The merchant scientist undertakes no personal financial risk. The focus on start-ups makes conflict of interest a serious concern within the NCE community. Interviewees worry about protecting the graduate students of merchant scientists, about the relation between research funding and company equity, and about the ethics of profiting personally from public funds. Conflicts of interest at the management level have occurred in the past, with professional staff demanding and receiving a percentage of equity in spin-offs. Potential conflicts exist when board members are also network partners, when network members appropriate institutional resources, and when commercial results are exaggerated and performance is over-reported to justify a continuation of public funding.

Finally, the threat of the removal of public funding, by way of the fourteen-year cap, may be a policy error. Network scientists were initially impressed that government had 'got it right' with the NCE program, and their approval was accompanied by a high level of buy-in. But they were uniformly opposed to the fourteen-year cap, feeling that the cut-off was arbitrary and denied them the right to compete for funding on merit. Sunset and future sustainability were key issues in CGDN for the whole of Phase III. Under pressure of the cap, instead of providing a fertile climate for research and translation, the program focused the participants' energies on profit and survival. Furthermore, for network scientists, the focus on profit interfered with the 'serious fun' of doing science and belonging to the network. The 'fun factor' is important for many scientists. A majority of respondents considered the effort to replace (public) program funding with funding from private sources to be misplaced, believing that self-sufficiency was doubtful without federal support. If the trajectory of CGDN could be reduced to a cipher, it would read like this: Phase I = 'Foundation'; Phase II = 'Translation'; Phase III = 'Speculation.'

Policy Focus

There is danger in focusing policy on scientific excellence and commercial relevance to the exclusion of other criteria. *Relevance is social as well as economic.* The economic interpretation, however, has been the predominant concern of the NCE program. At least within the biomedical networks, NCE funding underwrites the drive to production of products and processes that can be sold for profit. Until recently, relevance was considered to be synonymous with commodification. Even now, when the program's senior bureaucrats are beginning to factor improved population health into their economic equations, the contribution of network research to that goal is still seen in commodity terms – what one informant calls 'the search for a better pill.'

But, assuming a counterfactual universe, what if the $2 billion-plus allocated to the NCEs had been spent on programs to change people's behaviour, thereby reducing risk factors? Or on redistributing income and services to alleviate the effects of poverty and low social status? Although these social solutions have no commercial value, they are equally 'relevant,' and research consistently shows their effectiveness as determinants of human health. By focusing on saleable products, such as drugs and genetic tests, policy makers may be missing simple and cost-effective measures that can improve overall public well-being. Specifically, in terms of CGDN, claims of broad relevance may have obscured the point that genetic factors are only a limited component of the overall 'web of causation' in complex diseases.

Furthermore, a focus on 'excellence' may be less productive than a nurturing of 'hybrid vigour.' For much of CGDN's history, 'excellence' was interpreted so as to create exclusionary admission criteria that limited the variety of research and researchers recruited. Although these limitations produced organizational coherence and a strong core identity, they also constrained vitality by limiting the cross-fertilization that can occur when borders are more open. Hybrid vigour is further constrained when a single-minded focus on commodities and commercial relevance limits discussion and the sharing of results with 'outsiders.' Vigour clearly cannot be derived from the few 'strong ties' within an organization; it requires a diversity of 'weak ties' with multiple outsiders.[6]

Epilogue: Protecting the Public Interest

NCEs draw and defend their epistemic boundaries in various and unique ways. In one sense, the boundaries are removed. 'We are a nation of scientists. A community of scientists. We are everywhere and nowhere,' as Michael Hayden announced. In another sense, the boundaries are narrowly commercial and closely defended, policed by legally binding undertakings of confidentiality and non-disclosure. This book began with an assumption that academic science was turning away from *disinterested* enquiry and the open sharing of results and towards commercially *interested* enquiry and 'secret knowledge.' A second assumption was that the primary carriers of this change were new network forms of organization that crossed sectors and institutions and competed with traditional academic structures. The study reported here supports the assumptions, but with important qualifications. Network forms of organization were indeed ubiquitous, but traditional structures were embedded within them. Commercial interests and norms of (non-) disclosure were certainly a factor, but historical precedent argued against the novelty of the problem. Academic science has had a long and fruitful tradition of partnerships with industry in pursuit of useful products. What is new, however, is the assiduous policing of boundaries between ownership of knowledge and public rights of access, as well as exacerbated tensions between norms of commercial confidentiality and academic openness.

The research was positioned within the historically contingent distinctions between 'public' and 'private' domains and 'basic' and 'applied' science. Its larger purpose was to question the impact on 'the public interest' of shifts in the organization and the ethos of science.

The study suggests that concerns about the status of the public interest under the current policy regime are valid. The conversion of university research into marketable technologies is accelerating, despite the absence of informed public debate. Five categories of public interest concerns can be identified.

1 Strategic science policy measures rely on market ideologies. These measures may contribute to national prosperity in the short term, but over time they may prove problematic by contributing to fundamental shifts in the public/private divide. But because market ideologies are hegemonic, science, academy, industry, and state are all aligned along the same economic axis. Few formal institutions stand outside the system of market relations to speak for the public's long-term interests.[7]
2 The short-termist economic calculus of strategic science policy redirects funding towards commercially relevant research. Because academic science is funding-dependent, resources directed to discovery-based (non-commercial) research may be depleted. Yet this is the research that feeds the pool of public knowledge. Unless replenished, the pool may evaporate, leaving little for future researchers to build on or draw from.
3 The costs of building the public knowledge base have been socialized by decades of public support. But now, as the investment begins to pay off, the benefits are being privatized. Taxpayers pay twice: first to fund the initial research, then to buy it back in the form of proprietary knowledge.
4 The understanding of science as a public good is an important element of the culture of academic research. If science is redefined as proprietary, scientists' self-interest may come to predominate, with resulting negative effects.
5 Not only is the transfer of public assets and institutions to the private sector largely irreversible, it is also proceeding in the absence of public scrutiny and informed consent. The process may lack legitimacy.

In summary: the current science policy regime attempts a fundamental realignment of the public/private divide. The probable consequences will be far-reaching. Yet the realignment process seems largely ad hoc; meaningful analyses of societal costs and benefits are conspicuously absent. Given the potential significance of the problem, it is important to enquire into the dimensions of the impact on the public interest.

The current study of Networks of Centres of Excellence suggests, however, that it would be a mistake to overstate the extent of privatization in this case or the threat to the public interest that privatization represents. Despite attempts to redirect the culture and values of basic science, 'settlers' still predominate in CGDN. As the older generation retires, however, the balance may shift.

This study of NCEs has not definitively determined whether policy makers should mount a defence of the values of open science or continue to promote the networked integration of academic science and industry. The evidence suggests, however, that a hybrid of the two might be indicated. 'Open networks' could combine the best features of both worlds. They would provide a structurally flexible form consistent with the network model. But the basic and translational knowledge produced in these networks would not be proprietary – a principle consistent with open science. Practically, this would result in delegating further, pre-competitive development to an arm's-length, non-profit entity.

But, as argued earlier, it is somewhat misguided to posit a prior pristine status for 'open science' in its natural state. As has been asserted throughout this study, the public/private and basic/applied distinctions have always been fuzzy categories, constructed in action. The framing and funding of research agendas and the interpretation and application of scientific results are contingent and negotiated political achievements, conditioned by the interplay of power relations, market forces, social dynamics, and discursive strategies.

Still, since networks are becoming the default institutional structures in which public science is performed, and since public scientists are undertaking more and more adventures in the nature of trade, accountability is a valid concern. Policy makers perhaps need to reassess and reassert their expectations of what is properly open and properly closed, whether in public or private science.

Future Research

There is a clear need for further research in a number of areas related to the pursuit of relevance in Canada's public science system. First, further research is necessary to determine if the findings of this study can be replicated in other life-sciences networks. Findings derived from a single network covering a single discipline may be unique to that particular context. Data collected from scientists, administrators, and board members in other life-sciences networks could provide evidence that would validate, refute, or refine the findings and interpretations reported here. Further, comparisons with networks in other sectors would help distinguish the effects particular to the network form from those attributable to the disciplinary culture. Finally, international comparisons, particularly with the accumulated experience of the European research system, would help clarify what aspects are challenges specific to Canada rather than to science in general.

Second, further study is required to investigate 'translational research.' In the first case, a longitudinal study designed to investigate the trajectory of a discovery from the bench to the bedside and from the laboratory to the market would be especially useful, as would an assessment of the effects of 'transitional research' on fundamental enquiry. In the second case, study is needed to develop a means of measuring translational activities and including them in national research and development statistics. Canada has one of the lowest R&D:GDP ratios in the G8. The inclusion of translational activities would improve Canada's national research statistics.

Third, a detailed study of 'merchant science' is required to map the transition from the university to the market. Initially, there is a need to follow what happens to scientist founders as their companies scale up. Anecdotal evidence says that, beyond a certain stage, venture capitalists dilute founding shares to insignificance. Then, it would be useful to know what proportion of merchant scientists stay with their companies and resign their university positions as their companies grow. Also, we need to find a way to accurately measure the cost-benefit ratio of merchant science, taking into account opportunity costs as well as direct and indirect costs; benefits flowing back to the university, the network, and the state, in the form of dividends and taxation; and social returns and related cost savings in the form of new therapies and better health for Canadians. This would assist in identifying the appropriate level of overall public investment in merchant science.

Fourth, research is needed to explore the long-term impact of new institutional forms, such as NCEs, on existing research capacity in universities and hospitals. Some critics fear that the university is being 'hollowed out' as research increasingly moves to public-private institutes along the periphery.[8] What are the constitutional implications of redirecting funding from core university budgets, controlled by the provinces, to research organizations effectively controlled by the federal government?

Fifth, as discussed above, the results of this study indicate that, through the use of selective admission criteria and the regulation of access, the NCE is an intentionally elite program. Further research is needed to investigate the differential benefits accorded to scientists within the program, in comparison with those outside. Because scientists in NCEs have access to more resources than do other university faculty and because certain advantages accrue to scientists in these programs, there is a need to investigate whether NCEs are creating a scientific aristocracy at the expense of other researchers.

Finally, from the start, NCEs have been unsuccessful in integrating the social sciences into the program.[9] Only four of the thirty networks funded between 1990 and 2001 were concerned with the social sciences. In July 2001, two of the three remaining social science networks were cancelled. Research is needed to determine why the social sciences are not better represented in this elite program, both as integrated elements in other NCEs and as 'stand-alone' networks. An important question here is the returns in social reflexivity that could be added by the social sciences.

APPENDIX A

NCE Program Funded Networks, 1989–2005

	Sector[1]	Acronym	From	Headquarters Host
FUNDED NETWORKS				
Advanced Foods and Materials Network	HBT	AFMNet	2003	U. of Guelph
ArcticNet	NRI	ArcticNet	2003	U. de Laval
Automobile of the 21st Century	NRI	Auto21	2001	U. of Windsor
Canadian Language and Literacy Research Network	SS	CLLRNet	2001	U. of Western Ontario
Canadian Water Network	NRI	CWN	2001	U. of Waterloo
Stem Cell Genomics and Therapeutics Network	HBT	SCN	2001	Ottawa Hospital
Canadian Aquaculture Network	NRI	Aquanet	2000	Memorial U.
Canadian Stroke Network	HBT	CSN	2000	U. of Ottawa
Canadian Network for Vaccines and Immunotherapeutics	HBT	CANVAC	2000	U. de Montréal
Canadian Institute for Photonic Innovations	ETI	CIPI	1999	U. de Laval / York U.
Canadian Arthritis Network	HBT	CAN	1998	U. of Toronto
Geomatics for Informed Decisions	ETI	GEOIDE	1998	U. de Laval
Math of I.T. and Complex Systems	ETI	MITACS	1998	U. of Toronto
Intelligent Sensing for Innovative Structures	NRI	ISIS	1995	U. of Manitoba
Sustainable Forest Management Network	NRI	SFM	1995	U. of Alberta
Canadian Bacterial Diseases Network	HBT	CBDN	1989	U. of Calgary
Canadian Genetic Diseases Network	HBT	CGDN	1989	U. of British Columbia
Institute for Robotics and Intelligent Systems	ETI	IRIS	1989	PRECARN Inc.
Mechanical and Chemi-Mech Wood-Pulps	NRI	Wood-Pulps	1989	PAPRICAN
Microelectronic Devices, Circuits, Systems	ETI	Micronet	1989	U. of Toronto
Protein Engineering Network	HBT	PENCE	1989	U. of Alberta

APPENDIX A (concluded)

	Sector[1]	Acronym	From	Headquarters Host
NON-RENEWED NETWORKS				
Canadian Institute for Telecom. Research	ETI	CITR	1989–03	McGill U.
Health Evidence Application Linkage Net	ETI/SS	HEALNet	1995–02	McMaster U.
TeleLearning Research Network	ETI/SS	TL-RN	1995–02	Simon Fraser U.
Concrete Canada	NRI	Concrete	1989–98	U. of Sherbrooke
NeuroScience Network	HBT	Neuroscience	1989–98	Montreal Gen. Hospital
Respiratory Health Network	HBT	Inspiraplex	1989–98	Montreal Chest Hospital
Canadian Aging Research Network	SS	CARNET	1989–94	Concordia U.
Canadian Network for Space Research	NRI	CNSR	1989–94	U. of Western Ontario
Insect Biotech Canada	HBT	IBC	1989–94	Queen's U.
Molecular and Interfacial Dynamics C of E	HBT	CEMAID	1989–94	U. of Guelph
Ocean Production Enhancement Network	NRI	OPEN	1989–94	Dalhousie U.

Source: Compiled from NCE program documentation

1 This is not an 'official' classification; each network has been assigned to one of four broad sectors: NRI = Natural Resources and Infrastructure; HBT = Health and Biotechnologies; ETI = Electronics, Telecommunications, Information; SS = Social Sciences

APPENDIX B

DATA COLLECTION AND ANALYSIS

The majority of the data for this study were derived from in-depth interviews and participant observation, supplemented by an analysis of documents and financial and statistical reports. The preliminary phase of the study lasted from the autumn of 1998 through the spring of 1999. I collected and analysed documents, then interviewed CGDN officers in Vancouver and network researchers in Vancouver, Edmonton, and Calgary. At the same time, involvement in a separate study[1] of industry liaison offices (ILOs) in four universities (two in British Columbia and two in Alberta) allowed me to solicit information on network commercialization practices and network-university interface issues from the fifteen ILO officials I interviewed.

The next phase, extending from the autumn through to the end of 1999, focused on the federal level and the officials responsible for the NCE program. During a week-long fieldwork visit to Ottawa, I identified and interviewed a total of nineteen individuals[2] involved in the program's initiation, development, and ongoing maintenance. Historical details of policy formation and program building were sought, as well as the rationale behind certain 'design features,' such as the twin criteria of scientific excellence and commercial relevance. At the same time, documents and reports spanning the NCE's history and pre-history were collected from the program directorate. These materials included annual reports, program evaluations, public relations materials, newsletters, and various committee reports. Particular attention was paid to the acquisition of program-wide information on partnership and intellectual property arrangements, company creation, and funding patterns.

The final phase of data collection encompassed the CGDN case study, which extended from the end of 1999 through to the end of 2000, with follow-up visits to June 2001. CGDN was selected as the case-study network on the basis of the following criteria.

1 **Research sector** The case-study network had to be closely associated with the natural rather than the applied sciences because of the normative tensions associated with commercializing the natural sciences. For this reason, only networks in the health and biotechnologies (HBT) sector were considered.

2 **Position on the commercial continuum** Although all HBT networks tend to straddle the basic/applied divide, some are closer to commercialization than others. The goal was to select a network towards the basic end of the spectrum, because it was assumed that change would be perceived as more profound in this area than in others.
3 **Longevity** Because Phase I networks began operations in 1989–90 and were approaching 'sunset' on program funding, these more mature networks offered the best possibility of studying the program's entire trajectory.
4 **Density** A network with dense and diverse webs of researchers, partners, and collaborative arrangements would offer richer potential for exploring social capital and actor-networks than would one with a more limited range.
5 **Funding** The more resources a network commands, the more public and private tensions come into high relief. For these reasons, one of the 'wealthier' networks was selected.
6 **Convenience** Although networks are virtual and distributed, their activities are anchored largely on university and university-hospital campuses. For reasons of accessibility and economy, it was preferable to select a network headquartered at a nearby campus.

CGDN was the best match with these selection criteria.

In March 2000, I attended the network's annual scientific meeting in Vancouver, one of its key cultural events. The purpose was to present a paper introducing my study, and to conduct and solicit interviews, observe interactions, ask questions, and generally familiarize myself with the science and business of the network. Directly afterwards, I represented CGDN as a volunteer media relations officer at the International Human Genome Project's annual conference, which the network was co-hosting. These meetings were invaluable introductions to network culture and science and to the vast 'industry' that molecular biology has become. In addition, during the study, I made site visits to three research laboratories in Toronto and several to the Centre for Molecular Medicine and Therapeutics in Vancouver, for interviews and observation. But the core of my fieldwork centred on the network's cramped administrative headquarters in the 'NCE building' at the University of British Columbia. Here, for a period of eight weeks, I observed the workings of the network from a makeshift desk in the hallway.

Over the course of the study, I interviewed a selection of board members and private-sector partners and all current and former professional staff.[3] In selecting which of the more than fifty network researchers to interview, I focused on the 'founder population' – the sixteen scientists who remained active in the network of the twenty-one who had signed the original 1988 proposal. Eleven of the sixteen were interviewed. For balance, I also contacted two of the five founders who had left the network and three more recent recruits, two from the start of Phase II (1994–5), and another from the start of Phase III (1998–9). In total, the CGDN phase of data collection incorporated forty formal interviews with thirty-one people.

Interviews were semi-structured, allowing scope for reflection and opinion. Informants were first asked to describe their recollections of the network-building process, then answered a series of questions about the science produced in the network, culture and relationships, commercialization practices, governance, and whether they had noted any problems or 'sticking points' over the years. The relative weight of these questions was adjusted to reflect the informant's role in the network. The majority of interviews were conducted in Toronto, Ottawa, and Vancouver – three of the network's four main nodes. I was unable to visit Montreal, the fourth major centre, but interviewed two researchers from McGill, one by telephone and another during his visit to Vancouver.

Throughout the study, I attempted to compensate for the 'single-case' focus by identifying and interviewing other knowledgeable individuals with interests in the NCE program. These included 'insiders' involved in networks other than CGDN and 'outsiders' such as university technology managers and policy consultants. The purpose was to generate a cross-section of fact, opinion, and experience about NCEs from which shared patterns could emerge – patterns that would not be discernible in a single case.[4]

A total of seventy-four formal interviews were conducted with sixty-five people in nine Canadian and two U.S. cities (see table B.1). CGDN professional staff were interviewed twice, at the beginning and mid-point of the study, to check changing conditions and perceptions. Michael Hayden was interviewed three times: a wide-ranging discussion at the beginning of the study helped define my general focus; another at the mid-point dealt with the human genome program and the network's involvement in genomics; a third during fieldwork covered specific questions that had arisen and shifts I had noted. CGDN's

Table B.1
Formal interviews conducted

Interviewees	Number of people	Interviews
Senior policy makers	4	4
NCE directorate	7	7
NCE program officers	6	7
CGDN 'founder' researchers		
– Current	11	14
– Former	2	2
CGDN 'new' researchers	3	3
CGDN professional staff	4	9
CGDN private sector	5	5
Non-NCE scientific networks	3	3
University administrators	2	2
University technology managers	15	15
Policy consultants	3	3
TOTAL PEOPLE/INTERVIEWS	65	74

NCE program officer was interviewed twice; once in Ottawa, in October 1999, and again in Vancouver during the annual scientific meeting in March 2000.

Initial analysis of the data began during fieldwork. Daily write-up of field notes helped me to reflect on what I was discovering and to identify questions for subsequent follow-up. After fieldwork, during the intensive analysis of the data, I continued with the practice of daily written reflection. These notes reminded me of where my thinking had been in relation to the study and suggested directions I might explore. They proved invaluable in helping me structure the eventual write-up.

The materials I had collected included financial and statistical reports. My background as a professional accountant allowed me to analyse financial and performance data using generally accepted accounting principles and conventions. Key ratios were calculated in an attempt to determine the program's economic costs and benefits and comparative rates of public/private participation and reward. Such calculations are unable to account for social dimensions of the research questions, since social costs and benefits resist quantification. Nevertheless, these indicators can suggest the underlying social calculus, and it is in this spirit that they were sought.

The policy and program material was analysed and written up first. Several conference papers and articles were produced from these historical and interview data.[5] This process had the effect of 'stabilizing' a large part of the evidence. The 'macro' level of the program's composition and policy context could then be set aside in favour of a much finer-grained analysis of the network's micro-level practices. The material lent itself naturally to this bifurcation, leading me to question theoretical claims that actor-networks could not be bound within structural frames.

Next, network interviews were sorted into four broad categories: 'network scientists,' 'professional staff,' 'board and private sector,' and 'other.' Provisional codebooks were developed from iterative readings of the transcripts in each category, which were then coded and recoded using software tools of my own devising (rather than a commercial qualitative-analysis program). Once I was satisfied that the codings were consistent, each category was sorted by main code and subcodes. Then all categories were combined in a single database and sorted. A numerical weight was assigned to the codes according to frequency across categories. The dominant codes became headings to which less frequent codes were assigned on the basis of 'family resemblances.' These then provided a framework to guide the structure of the dissertation. In turn, these dominant codes were collapsed into broad interpretive themes, to aid theory building.

Notes

Preface

1 NCE (2002).
2 Clark (1998) is one exception, providing a comparative but theoretical overview of various 'formal knowledge networks.' As well, Robert Dalpé's ongoing research program includes interest in certain NCEs; for example, see Dalpé and Ippersiel (2000) and Dalpé et al. (2001).
3 A full description of data collection techniques and methods of data analysis can be found in appendix B.
4 A definitive source of information on qualitative research methods is Miles and Huberman (1994).

Introduction

1 Discussions can be found, for example, in Slaughter (1998), Slaughter and Leslie (1997), Gibbons et al. (1994), Rip (2000, 2001), in the work of Henry Etzkowitz and colleagues (1997, 1998, 2000), and in my own work: Atkinson-Grosjean (2002, 2003a), Atkinson-Grosjean et al. (2001), and Fisher et al. (2001).
2 An example often cited is the speculative purchase of a painting by someone other than an art dealer, with the intention of eventual resale at a profit.
3 See Atkinson-Grosjean (2002) and Rip (2000).
4 Consult Gordon (1991) and Rose (1996) on the 'enterprise self'; Hood (1991, 1995) for a full accounting of NPM more generally; Savoie (1995) for NPM's influence in Canada; and Power (1995) on accounting models.
5 Rip (2001: 4).
6 Later collapsed into the even more focused Industry Canada.

7 E-mail, 21 January 2000.
8 See the information commissioner's reports: Reid (2000, 2001).
9 This is not an isolated case. Another doctoral researcher attempted to explore a similar topic in Ottawa during the early 1990s. Claire Polster was seeking financial and statistical information on the proliferation of federal support programs for industry-relevant university science; some of the data required for her study were denied to her. Other data were not tracked, and what was tracked often proved inconsistent and unreliable (Polster 1993). Yet another researcher spent a year attempting to gain access to Health Canada without success (personal communication).
10 E-mail, 31 March 1999.
11 E-mail, 17 May 2000.
12 E-mail, 12 June 2000.
13 NCEs thus pioneered the model for subsequent large-scale funding initiatives in genomics. Genome Canada is incorporated as a private company governed by a board of directors, but 90 per cent funded from public programs. Accountability concerns are profound.

1 Two Divides

1 For Pasteur's Quadrant, see Stokes (1997). For the other formulations, see, respectively, Godin (2000–3), Callon (2002), and Holton and Sonnert (1999).
2 By Norberto Bobbio (1989) and Jeff Weintraub (1997).
3 Starr (1988).
4 The classic case is *Moore vs Regents of the University of California*; see Boyle (1996). John Moore sued researchers and their university for stealing his cell line (uniquely resistant to hairy-cell leukemia) for profit and without his consent. He lost the case.
5 Starr (1988: 2).
6 Arendt (1959: 56–7).
7 Huff (1997: 28).
8 Ibid., 33.
9 Shapin (1995a: 269). See also Shapin (1994) for further descriptions of 'gentlemen-scientists' like Robert Boyle.
10 Habermas (1989). For an interesting discussion on historiographical approaches to the relation of public sphere and private life, see Goodman (1992). A definitive critique of the inadequacy of the liberal model of the public sphere described by Habermas is available in Fraser (1997).
11 The key source is Arendt (1959), but see also Godin (2000–3: 3) for an interesting interpretation.

12 See Toulmin (1990) and (2001). Latour (1993) offers a parallel discussion.

13 Thanks to the late Stephen Straker for this point, and many other shafts of enlightenment.

14 Quoted in Kline (1995: 194).

15 Lenoir (1998) is a good discussion of late-nineteenth-century pharmaceutical initiatives. David Noble (1977) provides the definitive historical account of engineering in the United States. Charles Weiner (1986, 1989) and Roger Geiger (1988, 1990) discuss patenting issues, as does Noble. Thorstein Veblen (1918) is a prescient disquisition on the relationship between universities and commerce.

16 Kline (1995) provides a detailed account of the basic/applied distinction. See Julian Huxley (1934) for the first coinage of 'basic research.'

17 For the quoted sentence see Polanyi (1962: 62). Further discussion can be found in Polanyi (1940) and Sheehan (1993). See David (1995) for Bernal and Polanyi's legacy.

18 Geiger (1990: 19).

19 See Bush (1945: 19). For extensive discussions on science policy and the nature of the social contract, consult David Guston's work: for example, Guston (2000a) and Guston and Kenniston (1994). For the Canadian context, see chapter 2.

20 Godin (2000–1: 9) is the source for the concept of the autonomous researcher and the quotation on peer review and government. See Holton and Sonnert (1999: 53) for a discussion of the charter for pure science. The quotation from 'Endless Frontier' may be found in Bush (1945: 7).

21 Stokes (1995, 1997).

22 On Pasteur's Quadrant, see Stokes (1995, 1997). A similar formulation, found in Holton and Sonnert (1999), adopts 'Newtonian Science,' 'Baconian Science,' and 'Jeffersonian Science' as the ideal types. The latter emphasizes the role of state patronage in promoting scientific advance. For the two quotations, see Stokes (1995) pages 5 and 6, respectively.

23 Stokes had a long and distinguished career in U.S. science policy. He died of leukemia shortly after *Pasteur's Quadrant* went to press. Work has continued in Branscomb, Kodama, and Florida (1999), Nelson (1996), Nelson and Romer (1998), Holton and Sonnert (1999), Branscomb, Holton, and Sonnert (2000), and Sonnert and Brooks (2000).

24 For Canadian thinking on this issue, see, for example, Buchbinder (1993), Polster (1998), and Newson (1998). For the United States, see Sheila Slaughter and colleagues at the University of Arizona: for example, Slaughter and Leslie (1997), Slaughter (1998), and Slaughter and Rhoades (1990). Simon Marginson (1997) is a good source for Australia. On socialization of costs,

see Noble (1997). The ethics of commercialization are covered in, for example, Goldman (1989).

25 Dasgupta and David (1994).

26 Starr (1988: 2).

27 For the original positions, see Nelson (1959) and Arrow (1962). Keith Pavitt (2000) offers an updated interpretation.

28 Nelson (1998), Nelson and Sampat (2001). For an extended discussion on the economic costs and benefits of patents, see Mazzoleni and Nelson (1998).

29 Nelson (1998: 2).

30 See Cohen, Nelson, and Walsh (1996) on the efficiency of traditional diffusion models and Nelson and Romer (1998: 59) on dismantling barriers to access.

31 Florida and Cohen (1999: 590).

32 Rappert and Webster (1997); see also Packer and Webster (1995, 1996), Webster and Packer (1995, 1996), and Webster and Packer, eds (1996).

33 See Nelson and Romer (1998), Cohen et al. (1998), Blumenthal et al. (1996).

34 Heller and Eisenberg (1998).

35 Blumenthal et al. (1997).

36 Rahm (1994).

37 Nelson (1996: 141).

38 Ibid., 137. See also Nelson and Romer (1998).

39 House Committee on Science (1998: 57). This document was endorsed by the House of Representatives in October 1998 as a framework for an updated national science policy for the new century.

40 Quoted in *American Institute of Physics Bulletin of Science Policy News* 148 (28 October 1998).

41 Quotations are from Nelson and Romer (1998: 45). See also Rosenberg (1998) on improvement versus innovation.

42 David (1995: 13).

43 Callon (2002: 280).

44 Callon (1994: 401). The economic details of the arguments are beyond the scope of this book but are fully articulated in Dasgupta and David (1994); also David (1998a, 1998b, 2000), on the one hand, Callon (1994, 1998, 2002), and Callon, ed. (1998), on the other.

45 For additional discussion on this point, see Cambrosio and Keating (1998).

46 Callon (1994: 416).

47 Callon (2002: 281).

48 For the full exposition of Mode 1 and Mode 2, see Gibbons et al. (1994) and Nowotny, Scott, and Gibbons (2001); see also Jacob (2000) for an excellent summary.

49 Etzkowitz (1997: 149–50). For additional iterations of the model, see Etzkowitz, Webster, and Healey (1998) and Etzkowitz and Leydesdorff (1997).
50 Fuller (2000: xiii).
51 For 'fashionable ideas' and their features, see Rip (2000). Jacob and Hellstrom (2000) offer programmatic examples.
52 For the development of Rip's thinking over time, see Rip (1990, 1997, 2000, 2001) and van der Meulen and Rip (1996).
53 The following is derived from Whyte (1997), who relies on Ruivo (1994).
54 Marginson (1997: 73).
55 Gordon (1991) and Rose (1996) are sources for the 'enterprise self.' See Hood (1991, 1995) for a full accounting of NPM more generally, and Savoie (1995) for its influence in Canada. Power (1995) is the definitive source on accounting and auditing technologies.
56 Rip (2001: 4); see also van der Meulen and Rip (1996: 346–7).
57 Rip (2000) develops the notion of 'fashionable ideas.' Atkinson-Grosjean (2002) addresses the concept of 'international ideas.'
58 See Goldstein and Keohane (1994) for a full explanation of world views and principled and causal beliefs. A later chapter describes epistemic communities of scientists, but the term was first used in relation to the international policy community. See, for example, Ruggie (1975) and Haas (1992). Martha Finnemore's work is important in the area of 'teaching' science policy to member states; see Finnemore (1992, 1993). In terms of convergence at the OECD, a dialectic is at work in that many of the policy professionals are seconded from member states. According to an informed observer, a mutual shaping of policy occurs between and among the member countries, the Permanent Secretariat, and the expert communities.
59 For the quoted sentence, see Molas-Gallart and Salter (2001: 5). Merton (1973 [1968]) is the source for 'the Matthew effect in science.'
60 See Callon (2002) as well as Callon (1994), Nelson and Winter (1982), and Rip (1997).
61 Stokes (1995). The opinion belongs to Bernard Cohen, who was responding to Stokes's presentation, and responses are appended to the document. Cohen was discussing the Bowman Report, the foundation document for Bush's (1945) landmark study, *Science: The Endless Frontier.*
62 Critics of the changing milieu of academic knowledge production view phenomena such as patenting and public-private research partnerships as evidence of the intrusion of global capital and market ideologies into academic institutions. But in practice-based approaches, as Knorr-Cetina (1995) admits, these wider concerns disappear. See Winner (1993), Kleinman (1991, 1998), and Fuller (1992) for more on 'structural neglect.' Shapin

(1995b) talks about 'why' questions. Knorr-Cetina and Mulkay (1983: 6) discuss micro-level relational approaches. Latour (1987) issues the injunction to 'follow the actors.'

63 Keating and Cambrosio (2000: 385).

64 John Law describes these networks as 'punctualized'; see Law (1992: 385). It is well to remember, though, that even 'structural' networks can decompose, revealing their contingent natures.

65 Shapin (1995b: 293). For another perspective on 'studying up,' see Bronwyn Parry's (1998) interesting account of her attempt to study 'élite networks' of senior executives in big pharma and biotechnology.

66 Latour and Woolgar (1986: 27–8).

67 For the quoted sentences, see Latour and Woolgar (1986: 29, 30). On the concept of 'the stranger,' refer to Simmel (1950).

68 The science wars debate is well beyond the parameters of this book. For more information, see, for example, Koertge (1998), Segerstrale (2000), and others. Sokal and Fuller's quoted comments derive from interviews with the author for Atkinson-Grosjean (1997: 11–12).

69 Collins (2000); Collins and Evans (2002). For the purposes of this study, I set out to gain interactional expertise by immersing myself in readings about medical genetics and molecular biology, both prior to and during the study. I relied mainly on journals such as *Science*, *Nature*, and *Nature Genetics*. Although the 'empiricist repertoire' of the scientific sections of these journals was almost impossible to penetrate, the 'news' and 'features' sections were couched in an informal 'contingent repertoire' and proved much more accessible. (See Gilbert and Mulkay 1984 for an explanation of these dual repertoires.) I also tracked developments in the field by subscribing to electronic lists such as 'Medscape's Molecular Medicine MedPulse' and 'Science Week,' as well as activist monitors such as 'Genetic Crossroads' and 'Loka Alert.'

2 Science Policy in Canada and the NCE Experiment

1 The historical background presented in this chapter relies heavily on Phillipson (1983, 1991) but more especially on our personal correspondence. By virtue of his oral history projects in the 1970s and 1980s, Phillipson is an authority on the National Research Council and the evolution of Canadian science policy. He has communicated an enormous amount of background material to me in a series of letters over the period 1998–2001. His collegial willingness to share his scholarship has enriched my understanding, and I acknowledge his contribution to this policy history, which in many cases

draws directly on our correspondence. Parts of this chapter appeared in Atkinson-Grosjean, House, and Fisher (2001) and Atkinson-Grosjean (2002).

2 Most of these men were Canadian-born of British extraction, middle-class in origin, and Protestant. See John Porter (1965: 507–11) on power relations in Canada. For the operation of the U.S. power elite, see Mills (1956).

3 B. Lee, 'The Atom Secrets,' *Globe Magazine*, 28 October 1961; cited in John Porter (1965: 432).

4 University administrator, cited in Henderson (2001).

5 Phillipson (1991).

6 Thistle (1966: 26, 29, 127).

7 The enquiry was conducted by a parliamentary subcommittee struck in April 1919. The 'bargaining' quotation is attributed to Professor Lash Miller, University of Toronto, Cronyn Committee Proceedings, 4 June 1919: 99; cited in Lamontagne Committee (1968–77, vol. 1: 31).

8 Lamontagne Committee (1968–77, vol. 1: 61).

9 Massey Commission (1951: 175).

10 Ibid., 177.

11 Steacie (1965: 159–60). Steacie left McGill University to become head of NRC's chemistry division in 1939. He was appointed NRC's vice-president in 1950 and president in 1952, holding the latter post until his death in 1962, at which time he was widely acknowledged to be the 'leader of Canadian science' (Babbitt 1965: 3).

12 Lamontagne Committee (1968–77, vol. 1: 64).

13 Glassco Commission (1962–3, vol. 4: 230, 271).

14 Hayes (1973: 38–9).

15 OECD (1969: 63).

16 Among these, the Industrial Research Institute Program, established by the Department of Industry in 1966, provided grants to universities to establish institutes where they could work with industry and undertake contract research on their behalf. Legislative tools were also introduced: in 1967, government passed the Industrial Research and Development Incentives Act, which was intended to foster academy-industry collaboration in research aimed at solving industrial problems. As well, in 1969 the NRC announced a grants program for universities that emphasized the promotion of industrial development through 'centres of excellence' aimed at fostering a regional balance of scientific and technological expertise. However, plans for this program were vague.

17 Steacie (1965: 119), cited in Lamontagne Committee (1968–77, vol. 1: 269).

18 Lamontagne Committee (1968–77, vol. 1: 270–1).

19 Lindblom (1959) and Cohen, March, and Olsen (1972).
20 Dufour and de la Mothe (1993: 21, n13).
21 This remains a chronic problem. Only Italy has a lower ratio of R&D to GDP. The then finance minister, Paul Martin, made increasing the ratio a key commitment for the 2001–3 fiscal period.
22 Some of the material in this section also appeared in Atkinson-Grosjean (2002) and Fisher, Atkinson-Grosjean, and House (2001).
23 There were older initiatives. The Pulp and Paper Research Institute of Canada (Paprican), founded in 1925 at McGill University, represents perhaps Canada's most enduring example of a state-academy-industry alliance. Another enduring initiative is the NRC's Industrial Research Assistance Program (IRAP), launched in the 1960s, of which more will be said shortly. For the period in question, see Niosi (1995: 34–5).
24 For further discussion, see Friedman and Friedman (1990) and Niosi (2000).
25 Phillipson (1983).
26 Taped interview with Keith Glegg, January 2001.
27 See Callon (1997) and Callon, ed. (1998), for an analysis of the market significance of these networks; these should be read in relation to Granovetter's (1985) notion of 'embeddedness' in relation to economic action.
28 Thanks for this point go to the author's correspondent Andrew Russell, of the University of Colorado–Boulder, who is studying the development of computer research in the United States during the Cold War.
29 This imagery belongs to Janice Newson (1994).
30 See Clark (1998) for a descriptive account. Note that although NCE funds flowed to the networks through university financial systems, the university was only an intermediary. This remains the case in later iterations of the model, such as Genome Canada, where funds filter through a series of 'private' intermediaries before reaching the researcher. The university merely provides the final bank account in the series.
31 Pullen (1990).
32 Address reported in the NCE program internal newsletter *Liaison* 1 (1) (January 1990).
33 Brian Mulroney announced his resignation in February 1993. He stayed on as a caretaker prime minister until Kim Campbell won the leadership contest in June 1993. Their party was routed by the Liberals at the polls in October 1993, losing all but two seats.
34 Mazankowski's connection with NCEs lasted beyond his political career. In 2000, he became chair of CGDN's board of governors.
35 Industry Canada (1996: 9).
36 AUTM (1998). AUTM is the U.S.-based Association of University Technol-

ogy Managers. Since the majority of Canada's major research universities participate in the AUTM survey, these are fairly reliable indicators of growth for the 1990s. Conversely, the majority of Canadian universities do *not* participate in the AUTM survey, suggesting that commercialization concentrates in the major institutions, as in the United States. This is confirmed by Statistics Canada, which began its own survey of commercialization in the higher-education sector in 1998. Those figures show that the twelve most active universities account for 75 per cent of invention reports and licences, and two-thirds of new patent applications. Of the remaining universities, medium-sized institutions account for most of the activity (Statistics Canada 1999: 17). The number of universities that can effectively pursue commercialization activities and academy-industry partnerships thus appears limited.

37 Data are taken from Statistics Canada (1999) and preliminary analyses issued in April 2003 from the 2001 survey (Statistics Canada 2003a, 2003b).

38 For an idea of the relative significance of NCE funding, note that the three federal research councils were investing approximately $790 million per year in university research at this time (1997–8 per Statistics Canada. This figure includes NCE funding). The permanent NCE program was initially invested at $47.4 million annually but was quickly elevated to approximately $77 million. Thus NCE funding accounted for approximately 10 per cent of the total federal R&D contribution to university research.

39 NCE-SC (1997).

40 The first federally supported science initiative was the Geological Survey of Canada, founded in 1841, which laid the basis for the mining industry. In the 1880s, federal support of astronomy produced longitudinal maps, which were used in building the railways. The creation of experimental farms, patterned after the U.S. land grant movement, produced innovations suited to a cold climate and large gains in agricultural productivity. Before the end of the nineteenth century, several federal government departments had established national laboratories for the exploitation of natural resources.

3 Configuring the Canadian Genetic Diseases Network

1 I am using 'alignment' here to avoid the ANT term 'translation,' which means much the same thing: i.e., persuading potential allies that their interests and the network's interests are a good match. I want to retain 'translation' as much as possible for a later use that emerges from my empirical data – that of 'translational research.'

2 Callon (1986) and Callon and Latour (1981) provide good descriptions of the workings of power relations in networks. Cambrosio, Limoges, and Pronovost (1990: 214) deal with the way spokespersons create networks.
3 Knorr-Cetina (1999: 254).
4 CGDN-SCAN (1990–2000: 2).
5 I later discovered that 'civic science' was a buzzword at scientific conferences around this time. As he often did, Hayden had picked it up and adapted it for his own purposes.
6 The term belongs to Sabatier (1988).
7 Latour (1988) outlines the actor-network called 'Pasteur' and, derivatively, 'Hayden.'
8 Thanks to an anonymous reviewer for asking me to clarify my use of this term. It refers to the core group or intellectual elite in a research area or specialty. Harry Collins (1999: 163), for example, speaks of core-sets or core-groups of scientists 'who are at the heart of the knowledge-forming community in their specialist areas.' Thus CGDN's core-set, its founding intellectual elite, is also the intellectual elite of the discipline of medical genetics in Canada. Note that this meaning is somewhat distant from that of early controversy studies.
9 Wulf (1993).
10 Because of delays, the first phase lasted only a little more than three calendar years, although it spanned four fiscal years. The fiscal year ends on 31 March.
11 See CGDN-FP (1993); CGDN-NA (1994); CGDN-FP (1997); CGDN-NA (1998). Often, documents disagree. For example, the funding proposal will list more partner institutions than the network agreement does. When there was a discrepancy, the network agreements have been taken to be the more reliable source, since these list only formal signatories. However, the fact of the matter often lies in between.
12 This is one reason Industry Canada, in early planning for the program, wanted to insist on incorporation. As described earlier, the research councils resisted.
13 Lanzara (1983: 88).
14 In its February 2003 budget, the federal government committed to permanently fund the indirect costs of research carried by universities, create a program to improve graduate student training, and double the base budgets of the nation's granting councils. In return, Canadian universities agreed to double the amount of research and triple their commercialization output by the year 2010.
15 NCE-AR (Annual Report) (1996–7).

16 Comprising prior funding of fundamental research by the research councils, in the form of grants to network researchers for the basic element of their network research.
17 Merges (1996).
18 The NCE directorate requires such agreements. They govern all aspects of the relationships between a network and its university and industry partners.
19 Callon (1986: 205). See Nelson and Sampat (2001) and Atkinson-Grosjean and Fisher (1999) for a more thorough discussion of institutional constraints on ILOs.
20 In Canada, institutional policies differ. In about half of cases, the university owns; in the rest, the researcher owns. Even in the latter case, however, if the technology is to be developed, the university is assigned ownership rights and interests.
21 For an authoritative analysis of the former, see Lash and Urry (1994); for a Canadian perspective on the latter, see the articles in Holbrook and Wolfe (2000); also Wolfe (2000).
22 Murdoch (1995: 743).
23 Amin and Thrift (1994).
24 Kleinman (1998).
25 Lawrence and Locke (1997).
26 Knorr-Cetina (1999) found the same in her ethnography of a molecular biology laboratory.
27 Reported in *Chronicle of Higher Education* 23 February 2001.
28 Gibbons et al. (1994).
29 Callon (1999).
30 Latour (1998: 209).
31 Nowotny, Scott, and Gibbons (2001: 258–9).
32 Ibid., 258.
33 Gieryn (1995).
34 At the same time, the network stated on its website that 'no satisfactory policies will emerge if public concerns about genetics in health care are not addressed, and if those concerns are not fully and objectively researched.' See <www.cgdn.generes.ca>, accessed July 2001.
35 See, for example, the work of Kerr and colleagues (1997, 1998a, 1998b), Condit (2001), and Phillips and Orsini (2002).
36 Wynne (1999). Other examples include Epstein (1999), Haraway (1999), Irwin (2001), and Yearley (2000).
37 Kerr et al. (1997). See also Glasner (2000: 11). A troubling finding was that the scientists in this study failed to disclose their own social location and

vested interests while giving apparently objective assessments of the risks associated with the new genetics.

38 Wynne (1999).

39 Winner (1993: 369).

40 'Value for money' or 'comprehensive' audits are fundamental to New Public Management (Power 1995) and have now been adopted – at least in principle – by all federal and provincial auditors general.

41 For a fuller elucidation of contractarian approaches to science funding, see David Guston's recent work (e.g., 2000a, 2000b).

42 Guston (2000b: 33).

43 Foucault (1991 [1978]). The post-Foucauldian governmentality literature is extensive. But see, for example, Burchell, Gordon, and Miller (1991), Barry, Osborne, and Rose (1996), Power (1995), and Ericson and Haggerty (1997).

44 For more on accounting's 'calculative practices' and 'rituals of verification,' see Power (1995), T.M. Porter (1995), Miller (1994), and Harris (1998: 137) on surveillance and control.

45 In an international comparative study, Atkinson-Grosjean and Grosjean (2000) found that the proliferation of such intermediary agencies was a generalized feature of higher-education systems under neoliberalism.

46 Godin (2000–3: 16).

47 Figures are derived from my analysis of CGDN's annual financial and statistical reports CGDN-SR (1994–2000); CGDN-AR (1990–2001).

4 Culture and Science

1 Agre (1999). See also the classic discussions by Derek Price (1963) and Diane Crane (1972).

2 See Jean Lave and Etienne Wenger (1991) for an early contribution to the literature on communities of practice. Fleck (1979) proposes thought worlds, and Knorr-Cetina (1999) is the source for epistemic cultures in science.

3 CGDN-AR (1991: 8).

4 CGDN-FP (1993); emphasis in original.

5 CGDN-FP (2001).

6 CGDN-ASM (1991).

7 Details of industry partnerships appear in chapter 5.

8 The phrase 'practical science' was coined by historian R.G. Collingwood; these points were made by Evelyn Fox Keller (2000). For a political-economic perspective, see Mackenzie, Keating, and Cambrosio (1990).

9 Expert Panel Report; CGDN-EP (1997: 15). On priority issues in science, see Merton (1973 [1957]). Latour (1988) describes the 'extended lab' effect.

10 I am indebted to the grand rounds presentation of Prater and Newlands (1999) for much of the background in this section.
11 Keller (1995).
12 Childs (1999).
13 BIO (1989).
14 Richard Lewontin is an authoritative source (see 1991, 1999).
15 In Canada, note the work of Patricia Baird (e.g., 2000) and Clyde Hertzman (e.g., 1999). Both are members of CIAR.
16 Knorr-Cetina (1999).
17 See Gieryn (1999a, 1999b) for studies of scientific spaces.
18 Knorr-Cetina (1999).
19 Latour (1987).
20 Knorr-Cetina (1999).
21 Glasner and Rothman (1999) and Glasner (2000).
22 Within ANT, such recombinations are described as 'heterogeneous engineering.'
23 CGDN-EP (1992: 4).
24 Note that site visits assess all aspects of a network's mandate. In addition to the network's scientific program, its commercialization activities, partnerships and linkages, management, and training activities are also reviewed.
25 CGDN-EP (1993: 11).
26 CGDN-EP (1993: 8).
27 CGDN-EP (1993: cover letter).
28 CGDN-FP (1997: 11).
29 CGDN-EP (1997: 15, 17).
30 CGDN-AR (1999: 1).
31 CGDN-FP (2001: CD1).
32 CGDN-FP (2001).
33 NCE-SC (1997: 11).
34 Latour (1987) and Pickering (1993: 373).
35 Kohler (1998: 243).
36 See Pickering (1993: 374–5).
37 Kohler (1998: 249).
38 Law (1992).

5 From Science to Commerce

1 At the time of writing, a network of Canadian social scientists was engaged in a SSHRC-funded study ('Financing the Pipe') that explored the moral basis of profit when diseases are defined as a market opportunity (referred

to earlier in this book as 'profitable diseases'). The funding application (supplied to the author by the principals) contains powerful and evocative descriptive language, including the description just quoted.

2 Godin (2000–3: 7, n31).
3 Although the figures are much disputed, conventional estimates range between U.S.$300 million and U.S.$900 million to take a new drug through clinical trials and the regulatory approval process. The figure is lower or higher depending on whether the costs of marketing and public relations are included.
4 Michael Gresser, Merck Frosst director of chemistry, CGDN-AR (1991); emphasis added.
5 CGDN-FP (1988–S4).
6 CGDN-FP (1988: 3.7H).
7 CGDN-FP (1988: 3.6G.1).
8 E-mail, B. Lavers, 25 May 2000.
9 CGDN-FP (1993: 1.2).
10 CGDN-FP (1993: 1.8–9).
11 CGDN-FP (1997: 20).
12 Rip (1997: 635/10).
13 CGDN-FP (2001).
14 CGDN-EP (1997).
15 CGDN-FP (2001: 22). The main fundraising activity of the foundation is sponsoring of 'Jeans for Genes' day, from which approxately $500,000 was raised between 2001 and 2004.
16 PENCE-AR (2000).
17 Bush (1945: 83).
18 See Henderson (2001).
19 Excella made its first $500,000 investment in December 2000, in Vancouver-based Neuromed Inc. (CGDN-FP 2001).
20 The NCE program set up a 'Research Management Fund' (RMF) to provide transition assistance of up to $500,000 over one or two years to sustain networking activities for NCEs at the end of their funding window, but no additional subsidies were envisaged at the time of writing.
21 Rip (1997: 635/10).
22 Henderson (2001: 7).

6 Adventures in the Nature of Trade

1 In full: *Mystery, Company and Fellowship of Merchant Adventurers for the Discovery of Unknown Lands.* For an entertaining account from which the following details are drawn, see Milton (1999).

2 Milton (1999).
3 Rip (2000: 35–6). Rip relies on Eamon (1985); see also Jackson (2000).
4 Rip (2000: 35–6).
5 See Merton (1957), Gouldner (1957), and Glaser (1963), respectively.
6 Glaser (1963).
7 Kornhauser (1962).
8 Godin (2000–3).
9 Teaching classes, supervising students, and giving other types of service help reciprocate for the support received. But expectations in this regard are more collegial than administrative.
10 Zucker and Darby (1997: 503).
11 Slaughter and Leslie (1997). The term was originally coined in Hackett (1990).
12 Slaughter and Leslie (1997: 179).
13 Rip (2000).
14 Polanyi (1962).
15 An anonymous reviewer deserves a great deal of credit for helping me think through and clarify what I mean by the settler class of scientists.
16 This became quite apparent to this author during participant observation at CGDN's annual scientific meeting, in spring 2000. I observed the CEO playing the role of cheerleader, encouraging members to *start* thinking about the potential commercial applications of their research. For a ten-year-old network, with a mandate to commercialize, the exhortation struck me as telling. Few of the researchers at the meeting seemed to be in a commercial 'space.' The CEO might have been suggesting something novel. At the two-hour 'concurrent networking session,' four tables were set up. Session leaders dispensed information on bioinformatics, DNA sequencing, model organisms (knockout mice), and strategic funding for commercialization. Although the first three tables were extremely busy for the whole two hours, no one sat down to talk to the commercial director about the strategic funds the network makes available to move discoveries towards the market. *No one.* I kept checking back and noted that, eventually, the commercial director abandoned the table.
17 Fisher and Atkinson-Grosjean (2002), report on ILOs. Atkinson-Grosjean (2003b), reports on genome scientists.
18 The theoretical importance of 'translation' is well known in actor-network theory (ANT), which is also known as the 'sociology of translation.' In ANT, translation represents the way powerful actors enrol allies by aligning interests. This is not the sense it takes here, as the text will make clear.
19 See Gieryn (1995), on boundary work; Fujimura (1987), on articulation work.
20 The definition comes from a now-dead web page belonging to the U.S.

National Institute for Environmental Health. This covers most of the senses that have subsequently solidified.

21 See Birdsell, Atkinson-Grosjean, and Landry (2002).
22 NIH (2003).
23 It seems important to remember, though, that alternatives to market provision *are* available. Therapies could be developed and delivered entirely through public provision and the public health system. Funding could be invested in social and economic determinants of health rather than in new drugs and proprietary technologies. Commodification is a political choice, not a fact of nature.
24 For social worlds in sociological theory, see, for example, Becker (1982), Strauss (1982), Fujimura (1987, 1992), and Kaghan (1998).
25 The names assigned in this chapter are pseudonyms. However, given the nature of the network (there *is* only one CGDN), merchant scientists (there are only a handful), and career research programs (everyone knows who is working on what), it is impossible to fully disguise the identity of informants. One device used here is to add merchant voices from other life-sciences networks. Note also that members of the network's 'core-set' (e.g., Mike Hayden) gave permission for direct attribution.
26 Kohler (1998: 243).
27 Founded by Bob Korneluk and Alex MacKenzie of the University of Ottawa.
28 Launched by UBC's Mike Hayden and Frank Tufaro, in partnership with Max Cynader of the Neuroscience NCE.
29 Launched by Mike Hayden.
30 Brenda Gallie, of the Ontario Cancer Institute, and Kurt Vansande, a former economics professor from Simon Fraser University.
31 Frank Jirik, now of Calgary, and Chris Ong, of the University of British Columbia.
32 François Rousseau, Laval University.
33 Kathy Siminovich, Mount Sinai Hospital/University of Toronto.
34 Daniel Sinnett, Université de Montréal.
35 In one of the other life-sciences networks, for example, an officer was alleged to be running much of the network purchasing through his own companies. 'The specifics weren't as important as the clear indication that there were starting to be conflict-of-interest situations arising' (policy analyst).
36 CGDN-NA (1998: 37).
37 Recruited from the federal government. This person had been one of CGDN's program officers at the NCE directorate.
38 Shapin (1995b: 309).

39 The following two paragraphs appeared in Grosjean et al. (2000).
40 Cited in Bowe (1999: n1).
41 Bowe (1999: 1). The article relates to environmental values but provides an excellent introduction to the nature of values.
42 Bowe (1999: 4).
43 Baird (2003: section 6).

7 NCEs and the Public Interest

1 Star and Griesemer (1989).
2 Fujimura (1987).
3 Benoit Godin, personal communication.
4 For example, actor-network theory and Mode 2, or the spatial interpretations of Scott Lash and John Urry (Lash and Urry 1994; Urry 1998) and Manuel Castells (1996).
5 Mowery et al. (1999: 280).
6 Granovetter (1973).
7 The only non-market institutions left seem to be the law and the churches.
8 Gibbons (1998).
9 Continuing a tradition begun in Bush (1945), which also 'forgot' the social sciences.

Appendix B Data Collection and Analysis

1 'Academy-industry relations in North America,' Dr Donald Fisher, principal investigator, 1998–2001. Funded by SSHRC.
2 In the interim, the scope of the aforementioned SSHRC study had been extended to include NCEs, so this phase of my data-collection process overlapped with that of the larger study. Data from fifteen of the nineteen interviews were shared.
3 Access was controlled by staff; I was not permitted to contact board members and industry partners independently.
4 The technique, originally developed by Glaser and Strauss, is called 'maximum variation.' See Merriam (1998: 62) for a brief and useful description.
5 For example, Atkinson-Grosjean (1999a, 1999b, 1999c, 2000a, 2000b, 2002); Atkinson-Grosjean, House, and Fisher (2001); and Fisher, Atkinson-Grosjean, and House (2001).

References

Agre, Philip E. (1999). 'Visible colleges: infrastructure and institutional change in the networked university.' Department of Information Studies, UCLA. Draft paper dated 12 May. Accessed at http://dlis.gseis.ucla.edu/pagre/.

Amin, Ash, and Nigel Thrift (1994). 'Living in the global.' In Ash Amin and Nigel Thrift, eds, *Globalization, Institutions, and Regional Development in Europe*, 1–22. Oxford: Oxford University Press.

Anderson, Benedict (1983). *Imagined Communities*. London: Verso.

Arendt, Hannah (1959). *The Human Condition*. New York: Doubleday/Anchor.

Arrow, Kenneth (1962). 'Economic welfare and the allocation of resources for invention.' In Richard R. Nelson, ed., *The Rate and Direction of Inventive Activities*, 609–25. Princeton, NJ: Princeton University Press.

Atkinson-Grosjean, Janet (1997). 'Science wars: beyond the social text hoax.' In *C21: The World of Research at Columbia University, Special Section: The Sciences and the Humanities*. New York: Columbia University Press.

Atkinson-Grosjean, Janet (1999a). 'Excellence, networks, and the pursuit of profit: academic science and public policy in Canada.' Conference paper. San Diego (October): Society for the Social Studies of Science Conference.

Atkinson-Grosjean, Janet (1999b). 'Mapping the commercialization of university research in Canada.' Conference paper. Université de Sherbrooke (June): Congress of the Humanities and Social Sciences Federation of Canada.

Atkinson-Grosjean, Janet (1999c). 'Profits and loss: collaborations and commercialization at the public/private divide.' Conference paper. Reorganizing Knowledge: Transforming Institutions Knowing, Knowledge and the University in the XXI Century. University of Massachusetts at Amherst (September).

Atkinson-Grosjean, Janet (2000a). '"Adventures in the nature of trade": network science, Russian dolls, and a grand dichotomy.' Conference paper.

Demarcation Socialized: Or, How Can We Recognize Science When We See It? Cardiff University, Wales, August 2000.

Atkinson-Grosjean, Janet (2000b). 'CGDN and the NCE experiment: academic science at the public/private divide.' Conference presentation and poster presented at the Annual Scientific Meeting, Canadian Genetic Diseases Network, Vancouver, March.

Atkinson-Grosjean, Janet (2001). 'Adventures in the nature of trade: the quest for "relevance" and "excellence" in Canadian science.' PhD diss., University of British Columbia. November.

Atkinson-Grosjean, Janet (2002). 'Science and technology policy and university research: comparing Canada and the United States, 1979 to 1999.' *International Journal of Technology, Policy and Management* 2 (2): 102–24.

Atkinson-Grosjean, Janet (2003a). 'Canadian science at the public/private divide.' *Journal of Canadian Studies*, Special issue: Science and politics in Canada, 37 (3).

Atkinson-Grosjean, Janet (2003b). 'Politics, economics, ethics, and science: a study of Canadian genome scientists and funding organizations.' Invited presentation. Genomics Forum: Celebrating 50 Years of DNA Research. University of British Columbia and Genome British Columbia. 26 March.

Atkinson-Grosjean, Janet, and Donald Fisher (1999). 'Brokers on the boundary: academy/industry liaison in Canadian universities.' Conference paper. Society for the Social Studies of Science. San Diego, October.

Atkinson-Grosjean, Janet, and Garnet Grosjean (2000). 'The use of performance models in higher education: a comparative international review.' *Education Policy Analysis Archives* 8 (30), June: Accessed at http://olam.ed.asu.edu/epaa/v8n30.html.

Atkinson-Grosjean, Janet, Dawn House, and Donald Fisher (2001). 'Canadian science policy and public research organizations in the 20th century.' *Science Studies: An Interdisciplinary Journal for Science and Technology Studies* 14 (1): 3–25.

AUTM (Association of University Technology Managers) (1998). *AUTM Licensing Survey, Fiscal Year 1997*. Annual Survey of Member Institutions. Association of University Technology Managers, Inc. AUTM Licensing Survey.

Babbitt, J.D. (1965). 'Introduction.' In J.D. Babbitt, ed., *Science in Canada. Selections from the Speeches of E.W.R. Steacie*. Toronto: University of Toronto Press.

Baird, Patricia A. (2000). 'A genetic revolution in health and health care: policy challenges posed by new reproductive and genetic technologies.' Plenary paper: Science, Truth and Justice. CIAJ/ICAJ (Canadian Institute for the Administraton of Justice), Victoria, BC. October.

Baird, Patricia A. (2003). 'Getting it right: industry sponsorship and medical research.' *Canadian Medical Association Journal* 168 (10): 1267–9. 13 May.

Barry, Andrew, Thomas Osborne, and Nikolas Rose, eds (1996). *Foucault and Political Reason: Liberalism, Neoliberalism, and Rationalities of Government*. Chicago: University of Chicago Press.

Becker, Howard S. (1982). *Art Worlds*. Berkeley: University of California Press.

BIO (Biotechnology Industry Organization) (1989). *What Is Biotechnology?* Washington, DC: Biotechnology Industry Organization.

Birdsell, J., J. Atkinson-Grosjean, and R. Landry (2002). *Knowledge Translation in Two New Programs: Achieving 'the Pasteur Effect.'* Canadian Institutes of Health Research.

Blumenthal, D., N. Causino, E.G. Campbell, and K. Seashore Louis (1996). 'Relationships between academic institutions and industry in the life sciences – an industry survey.' *New England Journal of Medicine* 334: 368–73.

Blumenthal, D., E.G. Campbell, M.S. Anderson, et al. (1997). 'Withholding research results in academic life science: evidence from a national survey of faculty.' *Journal of the American Medical Association* 277 (16 April): 1224–8.

Bobbio, N. (1989). 'The great dichotomy: public/private.' In N. Bobbio, *Democracy and Dictatorship: The Nature and Limit of State Power.* Trans. Peter Kennedy. Cambridge, MA: MIT Press.

Bourdieu, P. (1975). 'The specificity of the scientific field and the conditions of the progress of reason.' *Social Science Information* 14 (6):19–47.

Bowe, Geoffrey S. (1999). 'Nature and value – some historical and contemporary reflections.' Paper published in Biopolitics International Organization conference proceedings, vol. 7, *Biopolitics: The Bioenvironment,* Budapest, 18–19 September 1998. Athens: Biopolitics International Organization.

Boyle, James (1996). *Shamans, Software and Spleens: Law and the Construction of the Information Society.* Cambridge, MA: Harvard University Press.

Branscomb, Lewis, Gerald Holton, and Gerhard Sonnert (2000). 'Science for society: cutting-edge basic research in the service of public objectives. A blueprint for an intellectually bold and socially beneficial science policy.' Conference report and consensus document. Conference on Basic Research in the Service of Public Objectives, Washington, DC, November. New York: Columbia University, Center for Science, Policy and Outcomes.

Branscomb, Lewis M., Fumio Kodama, and Richard Florida, eds (1999). *Industrializing Knowledge: University-Industry Linkages in Japan and the United States.* Cambridge, MA: MIT Press.

Buchbinder, H. (1993). 'The market oriented university and the changing role of knowledge.' *Higher Education* 26: 331–47.

Burchell, Graham, Colin Gordon, and Peter Miller, eds (1991). *The Foucault Effect: Studies in Governmentality.* Chicago: University of Chicago Press.

Bush, Vannevar (1945). *Science: The Endless Frontier.* Reprinted 1990. Washington, DC: National Science Foundation.

Callon, Michel (1986). 'Some elements of a sociology of translation: domestication of the scallops and the fishermen of San Brieuc bay.' In John Law, ed., *Power, Action, Belief: A New Sociology of Knowledge?* London: Routledge and Kegan Paul.

Callon, Michel (1994). 'Is science a public good?' *Science, Technology, and Human Values* 19 (4) (Autumn): 395–424.

Callon, Michel (1997). 'Actor-network theory: the market test.' Paper presented in absentia. Actor-Network and After. Keele University, U.K., July.

Callon, Michel (1998). 'An essay on framing and overflowing: economic externalities revisited by sociology.' In Michel Callon, ed., *The Laws of the Markets,* 244–69. Oxford: Blackwell.

Callon, Michel (1999). 'How concerned groups might be affected by their participation in scientific arenas: some lessons from the study of an association of patients.' Presidential plenary: The Participation of Lay People in the Production and Dissemination of Knowledge. Society for the Social Studies of Science (4S). San Diego, 27 October – 2 November.

Callon, Michel (2002). 'From science as an economic activity to socio-economics of scientific research: the dynamics of emergent and consolidated techno-economic networks.' In Philip Mirowski and Esther-Mirjam Sent, eds, *Science Bought and Sold: The Need for a New Economics of Science*. Chicago: University of Chicago Press.

Callon, Michel, ed. (1998). *The Laws of the Markets*. Oxford: Blackwell.

Callon, Michel, and Bruno Latour (1981). 'Unscrewing the big Leviathan: how actors macrostructure reality and how sociologists help them to do so.' In Karin D. Knorr-Cetina and Aaron V. Cicourel, eds, *Advances in Social Theory and Methodology: Towards an Integration of Micro- and Macro-Sociologies*. Boston: Routledge.

Cambrosio, Alberto, and Peter Keating (1998). 'Monoclonal antibodies: from local to extended networks.' In Arnold Thackray, ed., *Private Science: Biotechnology and the Rise of the Molecular Sciences*, 165–81. Philadelphia: University of Pennsylvania Press.

Cambrosio, Alberto, Camille Limoges, and Denyse Pronovost (1990). 'Representing biotechnology: an ethnography of Quebec science policy.' *Social Studies of Science* 20: 195–227.

Castells, Manuel (1996). *The Rise of Network Society.* Oxford: Blackwell.

CGDN-AR (1990–2001). Annual reports and financial statements of the Canadian Genetic Diseases Network. Vancouver: CGDN Archive.

CGDN-ASM (1991–2000). Annual scientific meetings: agendas and programs, Canadian Genetic Diseases Network. Vancouver: CGDN Archive.

CFDN-EP (1992). Report of the Expert Panel on the Canadian Genetic Diseases Network. Ottawa: Networks of Centres of Excellence Secretariat (accessed through CGDN Archive).

CGDN-EP (1993). Report of the Expert Panel on the Canadian Genetic Diseases Network. Ottawa: Networks of Centres of Excellence Secretariat (accessed through CGDN Archive).

CGDN-EP (1997). Report of the Expert Panel on the Canadian Genetic Diseases Network. Ottawa: Networks of Centres of Excellence Secretariat (accessed through CGDN Archive).

CGDN-FP (1988). Funding Proposal, Phase I, Canadian Genetic Diseases Network. Vancouver: CGDN Archive.

CGDN-FP (1993). Funding Proposal, Phase II, Canadian Genetic Diseases Network. Vancouver: CGDN Archive.

CGDN-FP (1997). Funding Proposal, Phase III, Canadian Genetic Diseases Network. Vancouver: CGDN Archive.

CGDN-FP (2001). Funding Proposal, Phase III, Mid-term Review, Canadian Genetic Diseases Network. Vancouver: CGDN Archive.

CGDN-NA (1994). Network Agreement, Phase II, Canadian Genetic Diseases Network. Vancouver: CGDN Archive.

CGDN-NA (1998). Network Agreement, Phase III, Canadian Genetic Diseases Network. Vancouver: CGDN Archive.

CGDN-SCAN (1990–2000). 'Scan' network newsletters, published sporadically. Canadian Genetic Diseases Network. Vancouver: CGDN Archive.

CGDN-SR (1994–2000). Annual statistical reports for Canadian Genetic Diseases Network, compiled for NCE Secretariat, Phases I and II. (No report prepared for 1996.) Vancouver: CGDN Archive.

CGDN-TR (1994). Transition Report: Phase I to Phase II, Canadian Genetic Diseases Network. Vancouver: CGDN Archive.

Childs, Barton (1999). 'Personal reflections on the history of medical genetics in the U.S.' Lecture in the series Human Genetics Past and Present, III. New York Academy of Medicine, 21 April.

Clark, Howard C. (1998). *Formal Knowledge Networks: A Study of Canadian Experiences.* Winnipeg: International Institute for Sustainable Development.

Cohen, M., J.G. March, and J. Olsen (1972). 'A garbage can model of organizational choice.' *Administrative Science Quarterly* 17: 1–25.

Cohen, Wesley M., R. Florida, L. Randazzesa, et al. (1998). 'Industry and the academy: uneasy partners in the cause of technological advance.' In Roger G. Noll, ed., *Challenges to Research Universities*, 171–99. Washington, DC: Brookings Institution Press.

Cohen, Wesley M., Richard R. Nelson, and John Walsh (1996). *Links and Impacts: New Survey Results on the Impact of University Research on Industrial R&D*. Pittsburgh: Department of Social and Decision Sciences, Carnegie Mellon University.

Collins, H.M. (1999). 'Tantalus and the aliens.' *Social Studies of Science* 29: 163–97.

Collins, H.M. (2000). 'Certainty and expertise in public domain science.' President's Lecture, University of British Columbia, Vancouver, BC. 14 September.

Collins, H.M., and Robert Evans (2002). 'The third wave of science studies.' *Social Studies of Science* 32: 235–96.

Condit, C. (2001). 'What is "public opinion" about genetics?' *Nature Reviews: Genetics 2001* 2: 811–15.

Crane, Diane (1972). *Invisible Colleges: Diffusion of Knowledge in Scientific Communities*. Chicago: University of Chicago Press.

Dalpé, Robert, Louis Bédard, and Marie-Pierre Ippersiel (2001). 'Interaction between science and technology in biotechnology – the case of photodynamic therapy.' Paper presented at Second Collaboration in Science and Technology (COLLNET) Conference. New Delhi, India. February.

Dalpé, Robert, and Marie-Pierre Ippersiel (2000). 'Public research organizations in the knowledge infrastructure.' In J. Adam Holbrook and David A. Wolfe, eds, *Innovations, Institutions, and Territory: Regional Innovation Systems in Canada*, 67–92. Innovation Systems Research Series. Kingston, ON: School of Policy Studies, Queen's University.

Dasgupta, P., and P. David (1994). 'Towards a new economics of science.' *Research Policy* 23: 487–522.

David, Paul A. (1995). 'Science reorganized? Postmodern visions and the curse of success.' Revised text version of speech. International Symposium on Measuring the Impact of R&D. Ottawa, 13–15 September.

David, Paul A. (1998a). 'Common agency contracting and the emergence of "open science" institutions.' *American Economic Review* 88 (2): 15–21.

David, Paul A. (1998b). 'Communication norms and the collective cognitive performance of "invisible colleges."' In G. Barba Navaretti, P. Dasgupta, K.-G. Maler, and D. Siniscalco, eds, *Creation and Transfer of Knowledge: Institutions and Incentives*, 115–66. Berlin: Springer.

David, Paul A. (2000). 'A tragedy of the public knowledge "commons"? Global science, intellectual property, and the digital technology boomerang.' *Electronic Journal of Intellectual Property Rights (Intellectual Property Research Centre, Oxford University)*. June.

Dufour, P., and J. de la Mothe (1993). 'The historical conditioning of S&T.' In P. Dufour and J. de la Mothe, eds, *Science and Technology in Canada*. London: Longman.

Eamon, William (1985). 'From the secrets of nature to public knowledge: the origins of the concept of openness in science.' *Minerva: Review of Science, Learning, Policy* 23 (3) (Autumn): 321–47.

Epstein, Steve (1999). 'New social movements and trends in the politics of knowledge production.' Presidential plenary: The Participation of Lay People in the Production and Dissemination of Knowledge. Society for the Social Studies of Science (4S). San Diego, CA, 27 October – 2 November.

Ericson, Richard V., and Kevin D. Haggerty (1997). *Policing the Risk Society.* Toronto: University of Toronto Press.

Etzkowitz, Henry (1997). 'The entrepreneurial university and the emergence of democratic corporatism.' In H. Etzkowitz and L. Leydesdorff, eds, *Universities in the Global Knowledge Economy,* 141–52. London: Cassell.

Etzkowitz, Henry, and Loet Leydesdorff (1997). 'Introduction.' In H. Etzkowitz and L. Leydesdorff, eds, *Universities in the Global Knowledge Economy,* 3–8. London: Cassell.

Etzkowitz, Henry, Andrew Webster, and Peter Healey, eds (1998). *Capitalizing Knowledge: New Intersections of Industry and Academia*. New York: SUNY Press.

Etzkowitz, H., E. Schuler, and M. Gulbrandsen (2000). 'The evolution of the entrepreneurial university.' In M. Jacob and T. Hellstrom, eds, *The Future of Knowledge Production in the Academy,* 40–60. Buckingham: Open University Press.

Finnemore, Martha (1992). 'Science, the state, and international society.' PhD diss. Stanford University.

Finnemore, Martha (1993). 'International organizations as teachers of norms: the United Nations Educational, Scientific, and Cultural Organization and science policy.' *International Organization* 47 (4) (Autumn): 565–97.

Fisher, Donald, and Janet Atkinson-Grosjean (2002). 'Brokers on the boundary: academy-industry liaison in Canadian universities.' *Higher Education* 44 (3–4): 449–67.

Fisher, Donald, Janet Atkinson-Grosjean, and Dawn House (2001). 'Changes in academy/industry/state relations in Canada: the creation and development of the Networks of Centres of Excellence.' *Minerva: A Review of Science and Policy* 39: 299–325.

Fleck, Ludwik (1979). *Genesis and Development of a Scientific Fact*. Foreword by Thomas S. Kuhn; ed. Thaddeus J. Trenn and Robert K. Merton; trans. Fred Bradley and Thaddeus J. Trenn. Chicago: University of Chicago Press.

Florida, Richard, and Wesley Cohen (1999). 'Engine or infrastructure? The university role in economic development.' In Lewis M. Branscomb, Fumio Kodama, and Richard Florida, eds, *Industrializing Knowledge: University-Industry Linkages in Japan and the United States*. Cambridge, MA: MIT Press.

Foucault, Michel (1991 [1978]). 'Governmentality.' In Graham Burchell, Colin Gordon, and Peter Miller, eds, *The Foucault Effect: Studies in Governmentality*, 87–104. Chicago: University of Chicago Press.

Fraser, N. (1997). 'Rethinking the public sphere.' In N. Fraser, *Justice Interruptus: Critical Reflections on the 'Postsocialist' Condition*, 69–98. New York: Routledge.

Friedman, Robert S., and Renee C. Friedman (1990). 'The Canadian universities and the promotion of economic development.' *Minerva* 28 (3) (Autumn): 272–93.

Fujimura, J.H. (1987). 'Constructing "do-able" problems in cancer research: articulating alignment.' *Social Studies of Science* 17: 257–93.

Fujimura, J.H. (1992). 'Crafting science: standardized packages, boundary objects and translation.' In Andrew Pickering, ed., *Science as Practice and Culture*. Chicago: University of Chicago Press.

Fuller, S. (1992). *Social Epistemology*. London: Routledge.

Fuller, S. (2000). 'Foreword.' In Merle Jacob and Tomas Hellstrom, eds, *The Future of Knowledge Production in the Academy*. Milton Keynes: Open University Press.

Geiger, Roger L. (1988). 'Milking the sacred cow: research and the quest for useful knowledge in the American university since 1920.' *Science, Technology, and Human Values* 13 (3 and 4) (Summer and Autumn): 332–48.

Geiger, Roger L. (1990). 'The American university and research.' In *The Academic Research Enterprise within the Industrialized Nations: Comparative Perspectives*. Report of a Symposium of the Government-University-Industry Roundtable. Washington, DC: National Academy of Sciences, 15–35.

Gibbons, M. (1998). 'Higher education relevance in the 21st century.' Paper prepared for UNESCO World Conference on Higher Education, Paris, 5–9 October 1998. Published by Secretary General, Association of Commonwealth Universities, supported by the World Bank.

Gibbons, M., C. Limoges, Helga Nowotny, Simon Schwartzman, Peter Scott, and Martin Trow (1994). *The New Production of Knowledge: The Dynamics of Science and Research in Contemporary Societies*. San Francisco: Sage.

Gieryn, Thomas F. (1995). 'Boundaries of science.' In S. Jasanoff, G.E. Markle,

J.C. Peterson, and T. Pinch, eds, *Handbook of Science and Technology Studies*. London: Sage.

Gieryn, Thomas F. (1999a). *Cultural Boundaries of Science: Credibility on the Line*. Chicago: University of Chicago Press.

Gieryn, Thomas F. (1999b). 'Truth-spots: the architectural emplacements of diverse verities.' Presentation in a lecture series. Scientific Ethos: Authority, Authorship and Trust in the Sciences. St John's College, University of British Columbia, Vancouver, 18 November.

Gilbert, Nigel, and Michael Mulkay (1984). *Opening Pandora's Box*. Cambridge: Cambridge University Press.

Glaser, B. (1963). 'The local-cosmopolitan scientist.' *American Journal of Sociology* 69: 249–59.

Glasner, P. (2000). 'Policy issues in genome research.' Conference paper presented at Demarcation Socialised: Or, How Can We Recognize Science When We See It? Cardiff, U.K., 24–8 August.

Glasner, P., and H. Rothman (1999). 'Does familiarity breed concern? Bench scientists and the human genome mapping project.' *Science and Public Policy* 26: 313–24.

Glassco Commission (1962–3). *Report of the Royal Commission on Government Organization*. 5 vols. Ottawa: Queen's Printer.

Godin, Benoit (2000–1). *Outline for a History of Science Measurement*. Paper No. 1. Project on the history and sociology of science and technology statistics. Montreal: Observatoire des sciences et des technologies, UQAM, 33 pages.

Godin, Benoit (2000–3). *Measuring Science: Is There 'Basic Research' Without Statistics?* Paper No. 3. Project on the history and sociology of science and technology statistics. Montreal: Observatoire des sciences et des technologies, UQAM, 29 pages.

Godin, Benoit (2001–6). *The Disappearance of Statistics on Basic Research in Canada: A Note*. Paper No. 6. Project on the history and sociology of science and technology statistics. Montreal: Observatoire des sciences et des technologies, UQAM, 30 pages.

Godin, Benoit (2001–7). *Defining R&D: Is Research Always Systematic?* Paper No. 7. Project on the history and sociology of science and technology statistics. Montreal: Observatoire des sciences et des technologies, UQAM, 17 pages.

Goldman, Alan H. (1989). 'Ethical issues in proprietary restrictions on research results.' In Vivian Weil and John W. Snapper, eds, *Owning Scientific and Technical Information: Value and Ethical Issues*, 69–82. New Brunswick, NJ: Rutgers University Press.

Goldstein, Judith, and Robert O. Keohane (1994). 'Ideas and foreign policy: an analytical framework.' In Judith Goldstein and Robert O. Keohane, eds, *Ideas and Foreign Policy*, 3–30. Ithaca, NY: Cornell University Press.

Goodman, Dena (1992). 'Public sphere and private life: towards a synthesis of current historiographical approaches to the old regime.' *History and Theory* 31 (1): 1–20.

Gordon, C. (1991). 'Governmental rationality: an introduction.' In G. Burchell, C. Gordon, and P. Miller, eds, *The Foucault Effect: Studies in Governmentality*, 1–52. Chicago: University of Chicago Press.

Gouldner, A. (1957–8). 'Cosmopolitans and locals: toward an analysis of latent social roles.' *Administrative Science Quarterly* Part I (3): 1957; Part 2 (4): 1958.

Granovetter, Mark (1973). 'The strength of weak ties.' *American Journal of Sociology* 78: 1360–80.

Granovetter, Mark (1985). 'Economic action and social structure: the problem of embeddedness.' *American Journal of Sociology* 91 (3): 481–510.

Grosjean, G., J. Atkinson-Grosjean, K. Rubenson, and D. Fisher (2000). 'Measuring the unmeasurable: paradoxes of accountability and the impacts of performance indicators on liberal education in Canada.' Second in a series of four studies prepared for the Humanities and Social Sciences Federation of Canada. June. Accessed at http://www.hssfc.ca/ResearchProj/PerfInd/FinalReportEng.html.

Guston, David H. (2000a). *Between Politics and Science: Assuring the Integrity and Productivity of Research*. Cambridge: Cambridge University Press.

Guston, David H. (2000b). 'Retiring the social contract for science.' *Science, Technology, and Human Values* 16 (4) (Summer): 32–7.

Guston, David H., and Kenneth Kenniston (1994). 'Introduction: the social contract for science.' In David H. Guston and Kenneth Kenniston, eds, *The Fragile Contract: University Science and the Federal Government*, 1–41. Cambridge, MA: MIT Press.

Haas, Peter (1992). 'Introduction: epistemic communities and international policy coordination.' *International Organization* 46 (1) (Winter): 1–35.

Habermas, Jurgen (1989). *The Structural Transformation of the Public Sphere: An Inquiry into a Category of Bourgeois Society.* Trans. Thomas Burger with Frederick Lawrence. Cambridge, MA: MIT Press.

Hackett, E.J. (1990). 'Science as a vocation in the 1990s: the organizational culture of academic science.' *Journal of Higher Education* (May–June): 241–79.

Hacking, Ian (1983). *Representing and Intervening: Introductory Topics in the Philosophy of Natural Science*. Cambridge: Cambridge University Press.

Hacking, Ian (1990). *The Taming of Chance*. Cambridge: Cambridge University Press.

Haraway, Donna (1999). 'For the love of a good dog.' Presidential plenary: The Participation of Lay People in the Production and Dissemination of Knowledge. Society for the Social Studies of Science (4S). San Diego, 27 October – 2 November.

Harris, J. (1998). 'Performance models.' *Public Productivity and Management Review* 22 (2) (December): 135–40.

Hayes, R. (1973). *The Chaining of Prometheus: Evolution of a Power Structure for Canadian Science*. Toronto: University of Toronto Press.

Heller, Michael A., and Rebecca S. Eisenberg (1998). 'Can patents deter innovation? The anticommons in biomedical research.' *Science* 280 (5364) (1 May): 698–701.

Henderson, Mark (2001). 'Life sciences commercialization initiative seeks to capture benefits of Canada's growing research base.' *Re$earch Money* 4 April.

Hertzman, Clyde (1999). 'Population health.' Seminar presentation. Interdisciplinary Studies Graduate Program. Green College, University of British Columbia, Vancouver, 3 March.

Holbrook, J. Adam, and David A. Wolfe, eds (2000). *Innovations, Institutions, and Territory: Regional Innovation Systems in Canada*. The Innovation Systems Research Series. Kingston, ON: School of Policy Studies, Queen's University.

Holton, Gerald, and Gerhard Sonnert (1999). 'A vision of Jeffersonian science.' *Issues in Science and Technology* 16 (1). www.issues.org/issues/16.1/holton.htm.

Hood, C. (1991). 'A public management for all seasons?' *Public Administration* 69 (Spring): 3–19.

Hood, C. (1995). 'The "new public management" in the 1980s: variations on a theme.' *Accounting, Organizations and Society* 20 (2/3): 93–109.

Hoskin, K.W. (1993). 'Accounting as discipline: the overlooked supplement.' In E. Messer-Davidov, D.R. Shumway, and D.J. Sylvan, eds, *Knowledges: Historical and Critical Studies in Disciplinarity*, 25–53. Charlottesville: University Press of Virginia.

House Committee on Science (1998). *Unlocking Our Future: Toward a New National Science Policy, A Report to Congress, September 24*. Washington, DC: U.S. Government.

Huff, Toby (1997). 'Science and the public sphere: comparative institutional development in Islam and the west.' *Social Epistemology* 11 (1): 25–37.

Huxley, Julian (1934). *Scientific Research and Social Needs*. London: Watts.

Industry Canada (1996). *Science and Technology for the New Century: A Federal Strategy*. Ottawa: Government of Canada.

Irwin, Alan (2001). 'Constructing the scientific citizen: science and democracy in the biosciences.' *Public Understanding of Science* 10: 1–18.

Jackson, Myles W. (2000). *Spectrum of Belief: Joseph von Fraunhofer and the Craft of Precision Optics*. Cambridge, MA: MIT Press.

Jacob, Merle (2000). '"Mode 2" in context: the contract researcher, the university and the knowledge society.' In Merle Jacob and Tomas Hellstrom, eds, *The Future of Knowledge Production in the Academy,* 11–27. Buckingham: Open University Press.

Jacob, Merle, and Tomas Hellstrom, eds (2000). *The Future of Knowledge Production in the Academy.* Buckingham: Open University Press.

Kaghan, William N. (1998). 'Court and spark: studies in professional university technology transfer management.' PhD diss., University of Washington, Seattle.

Keating, Peter, and Alberto Cambrosio (2000). 'Biomedical platforms.' *Configurations* 8: 337–87.

Keating, Peter, and Alberto Cambrosio (2003). *Biomedical Platforms: Reproducing the Normal and the Pathological in Late 20th Century Medicine*. Cambridge, MA: MIT Press.

Keller, Evelyn Fox (1995). *Refiguring Life: Metaphors of Twentieth-Century Biology.* New York: Columbia University Press.

Keller, Evelyn Fox (2000). 'Theory and practice in contemporary biology: epistemological cultures in science.' St John's College Speakers Series, University of British Columbia, Vancouver. Scientific Ethos: Authority, Authorship, and Trust in the Sciences, 16 March.

Kerr, A., S. Cunningham-Burley, and A. Amos (1997). 'The new genetics: professionals' discursive boundaries.' *Public Understanding of Science* 4: 243–53.

Kerr, A., S. Cunningham-Burley, and A. Amos (1998a). 'Eugenics and the new genetics in Britain: examining contemporary professionals' accounts.' *Science, Technology, and Human Values* 23 (3) (Summer): 175–98.

Kerr, A., S. Cunningam-Burley, and A. Amos (1998b). 'The new genetics and health: mobilizing lay expertise.' *Public Understanding of Science* 7: 41–60.

Kleinman, Daniel Lee (1991). 'Conceptualizing the politics of science: a response to Cambrosio, Limoges, and Pronovost.' *Social Studies of Science* 21: 769–74.

Kleinman, Daniel Lee (1998). 'Untangling context: understanding a university laboratory in the commercial world.' *Science, Technology, and Human Values* 23 (3) (Summer): 285–314.

Kline, Ronald (1995). 'Construing "technology" as "applied science." Public rhetoric of scientists and engineers in the United States, 1880–1945.' *Isis* 86: 194–221.

Knorr-Cetina, Karin (1995). 'Laboratory studies: the cultural approach.' In

S. Jasanoff, G.E. Markle, J.C. Peterson, and T. Pinch, eds, *Handbook of Science and Technology Studies*, 140–65. Thousand Oaks: Sage.

Knorr-Cetina, Karin (1999). *Epistemic Cultures: How the Sciences Make Knowledge*. Cambridge, MA: Harvard University Press.

Knorr-Cetina, Karin, and Michael Mulkay (1983). 'Introduction: emerging principles in social studies of science.' In Karin D. Knorr-Cetina and Michael Mulkay, eds, *Science Observed: Perspectives on the Social Study of Science*, 1–17. London: Sage.

Koertge, N. (1998). *Houses Built on Sand: Flaws in the Cultural Studies Account of Science*. Oxford: Oxford University Press.

Kohler, Robert E. (1998). 'Moral economy, material culture and community in drosophila genetics.' In Mario Biagioli, ed., *The Science Studies Reader*, 243–57. New York: Routledge.

Lamontagne Committee (1968–77). *A Science Policy for Canada.* Report of the Senate Special Committee on Science Policy. 3 vols. Ottawa: Queen's Printer.

Lanzara, Giovan Francesco (1983). 'Ephemeral organizations in extreme environments: emergence, strategy, extinction.' *Journal of Management Studies* 20 (1) (January): 71–95.

Lash, S., and J. Urry (1994). *Economies of Signs and Spaces*. London: Sage.

Latour, Bruno (1986). 'The powers of association.' In John Law, ed., *Power, Action and Belief: A New Sociology of Knowledge?*, 264–80. London: Routledge and Kegan Paul.

Latour, Bruno (1987). *Science in Action: How to Follow Scientists and Engineers through Society.* Cambridge, MA: Harvard University Press.

Latour, Bruno (1988). *The Pasteurization of France*, with 'Irreductions.' Trans. Alan Sheridan and John Law. Cambridge, MA: Harvard University Press.

Latour, Bruno (1992). 'Where are the missing masses? Sociology of a few mundane artefacts.' In W. Bijker and J. Law, eds, *Shaping Technology, Building Society: Studies in Sociotechnical Change*, 225–58. Cambridge, MA: MIT Press.

Latour, Bruno (1993). *We Have Never Been Modern*. Cambridge, MA: Harvard University Press.

Latour, Bruno (1998). 'From the world of science to the world of research?' *Science* 280 (5361) (10 April): 208–9.

Latour, Bruno, and S. Woolgar (1986). *Laboratory Life: The Construction of Scientific Facts*. Reprint of 1979 edition, *Laboratory Life: The Social Construction of Scientific Facts*, with a new introduction and postscript, and retitled to exclude 'social.' Princeton, NJ: Princeton University Press.

Lave, Jean, and Etienne Wenger (1991). *Situated Learning: Legitimate Peripheral Participation*. Cambridge: Cambridge University Press.

Law, John (1992). 'Notes on the theory of the actor-network: ordering, strategy, and heterogeneity.' *Systems Practice* 5 (4): 379–93.

Law, John, and J. Hassard, eds (1999). *Actor Network Theory and After.* Sociological Review Monographs. Oxford: Blackwell.

Lawrence, Peter A., and Michael Locke (1997). 'Editorial.' *Nature* (24 April): 757–8.

Lee, N., and S. Brown (1994). 'Otherness and the actor-network.' *American Behavioral Scientist* 37: 772–90.

Lenoir, Timothy (1998). 'Revolution from above: the role of the state in creating the German research system, 1810–1910.' *American Economic Review* 88 (2).

Lewontin, Richard (1991). *Biology as Ideology: The Doctrine of DNA*. Toronto: Anansi.

Lewontin, Richard (1999). 'In the blood: biologizing the social.' St John's College Speakers Series, University of British Columbia, Vancouver. Scientific Ethos: Authority, Authorship, and Trust in the Sciences. 21 October.

Lindblom, Charles E. (1959). 'The science of muddling through.' *Public Administration Review* 19: 79–99.

Mackenzie, Michael, Peter Keating, and Alberto Cambrosio (1990). 'Patents and free scientific information in biotechnology: making monoclonal antibodies proprietary.' *Science, Technology, and Human Values* 15 (1) (Winter): 65–83.

Marginson, Simon (1997). *Markets in Education*. Sydney: Allen and Unwin.

Massey Commission (1951). *Report of the Royal Commission on National Development in the Arts, Letters, and Sciences, 1949–51*. Ottawa: Government of Canada.

Mazzoleni, Roberto, and Richard R. Nelson (1998). 'Economic theories about the benefits and costs of patents.' *Journal of Economic Issues* 32 (4) (December): 1031–52.

Merges, Robert P. (1996). 'Property rights theory and the commons: the case of scientific research.' In Ellen Frankel Paul, Fred D. Miller Jr, and Jeffrey Paul, eds, *Scientific Innovation, Philosophy and Public Policy,* 145–67. Cambridge: Cambridge University Press.

Merriam, Sharon B. (1998). *Qualitative Research and Case Study Applications in Education.* San Francisco: Jossey-Bass.

Merton, Robert K. (1957). *Social Theory and Social Structure*. Glencoe: Free Press.

Merton, Robert K. (1973 [1942]). 'The normative structure of science.' In Norman W. Storer, ed., *The Sociology of Science: Theoretical and Empirical Investigations*, 267–78. Chicago: University of Chicago Press.

Merton, Robert K. (1973 [1957]). 'Priorities in scientific discovery.' In Norman W. Storer, ed., *The Sociology of Science: Theoretical and Empirical Investigations*, 286–324. Chicago: University of Chicago Press.

Merton, Robert K. (1973 [1968]). 'Behavior patterns of scientists.' In Norman W. Storer, ed., *The Sociology of Science: Theoretical and Empirical Investigations*, 325–42. Chicago: University of Chicago Press.

Michael, Mike, and Lynda Birke (1994). 'Enrolling the core-set: the case of the animal experimentation controversy.' *Social Studies of Science* 24 (1) (February): 81–95.

Miles, Matthew B., and A. Michael Huberman (1994). *Qualitative Data Analysis*. 2nd ed. Thousand Oaks: Sage.

Miller, P. (1994). 'Accounting as a Social and Institutional Practice: An Introduction.' In A.G. Hopwood and P. Miller, eds, *Accounting as a Social and Institutional Practice*, 1–39. Cambridge: Cambridge University Press.

Mills, C. Wright (1956). *The Power Elite*. New York: Oxford University Press.

Milton, Giles (1999). *Nathaniel's Nutmeg: Or, the True and Incredible Adventures of the Spice Trader Who Changed the Course of History*. New York: Farrar, Straus and Giroux.

Molas-Gallart, Jordi, and Ammon J. Salter (2001). *Living with Mediocrity: A Comment on Research Excellence and Patented Innovation*. Science Policy Research Unit Working Paper. University of Sussex: Science Policy Research Unit.

Mowery, David, Richard R. Nelson, Bhaven N. Sampat, and Arvids A. Ziedonis (1999). 'The effects of the Bayh-Dole Act on U.S. university research and technology transfer.' In Lewis M. Branscomb, Fumio Kodama, and Richard Florida, eds, *Industrializing Knowledge: University-Industry Linkages in Japan and the United States*, 269–306. Cambridge, MA: MIT Press.

Murdoch, Jonathan (1995). 'Actor-networks and the evolution of economic forms: combining description and explanation in theories of regulation, flexible specialization, and networks.' *Environment and Planning A* 27: 731–57.

NCE (2002). Evaluation of the Networks of Centres of Excellence: Final Report. 26 June: 15.

NCE-AR (1990–2001). Annual Reports for the NCE Program. Ottawa: Networks of Centres of Excellence Secretariat.

NCE-SC (1997). Selection Committee Report, NCE Program, Phase III. Ottawa: Networks of Centres of Excellence Secretariat.

Nelson, Richard R. (1959). 'The simple economics of basic scientific research.' *Journal of Political Economy* 67: 297–306.

Nelson, Richard R. (1996). *The Sources of Economic Growth*. Cambridge, MA: Harvard University Press.

Nelson, Richard R. (1998). 'Technology transfer, in theory and practice.' *C21: The World of Research at Columbia University* 3 (1) (Spring).

Nelson, Richard R., and Paul M. Romer (1998). 'Science, economic growth, and public policy.' In Dale Neef, Anthony Siesfeld, and Jaqueline Cefola, eds, *The Economic Impact of Knowledge*, 43–60. Resources for the knowledge-based economy. Boston, MA: Butterworth Heinemann.

Nelson, Richard R., and Bhaven N. Sampat (2001). 'Making sense of institutions as a factor shaping economic performance.' *Journal of Economic Behavior and Organization* 44: 31–54.

Nelson, Richard R., and S. Winter (1982). *An Evolutionary Theory of Economic Change*. Cambridge, MA: Harvard University Press.

Newson, Janice A. (1994). 'Subordinating democracy: the effects of fiscal retrenchment and university-business partnerships on knowledge creation and knowledge dissemination in universities.' *Higher Education* 27: 141–61.

Newson, Janice A. (1998). 'The corporate-linked university: from social project to market force.' *Canadian Journal of Communication* 23 (1) (Winter): 107–24.

NIH (National Institutes of Health) (2003). 'NIH roadmap: accelerating medical discovery to improve health.' http://nihroadmap.nih.gov.

Niosi, Jorge (1995). *Flexible Innovation: Technological Alliances in Canadian Industry.* Montreal and Kingston: McGill-Queen's University Press.

Niosi, Jorge (2000). *Canada's National System of Innovation*. Montreal: McGill-Queen's University Press.

Noble, David (1977). *America by Design: Science, Technology, and the Rise of Corporate Capitalism*. New York: Alfred A Knopf.

Noble, David (1997). 'Digital diploma mills: the automation of higher education.' Accessed at http://www.journet.com/twu/deplomamills.html.

Nowotny, Helga, Peter Scott, and Michael Gibbons (2001). *Re-thinking Science: Knowledge and the Public in an Age of Uncertainty.* Cambridge: Polity Press in association with Blackwell Publishers Ltd.

OECD (Organization for Economic Co-operation and Development) (1969). *Reviews of National Science Policy: Canada.* Paris: OECD.

OECD (1998). *Science, Technology, and Industry Outlook 1998*. Paris: OECD.

OECD (2002–2). *Science, Technology, and Industry Outlook 2002*. Paris: OECD.

Packer, Kathryn, and Andrew Webster (1995). 'Inventing boundaries: the prior art of the social world.' *Social Studies of Science* 25: 107–17.

Packer, Kathryn, and Andrew Webster (1996). 'Patenting culture in science: reinventing the scientific wheel of credibility.' *Science, Technology, and Human Values* 21 (4) (Fall): 427–53.

Parry, Bronwyn (1998). 'Hunting the gene-hunters: the role of hybrid networks, status, and chance in conceptualizing and accessing "corporate elites."' *Environment and Planning A* 30 (12) (December): 2147–62.

Pavitt, Keith (2000). *Public Policies to Support Basic Research: What Can the Rest of the World Learn from US Practice? (And What They Should Not Learn)*. Science Policy Research Unit Working Paper 53. University of Sussex: Science Policy Research Unit, 25 pages.

Pels, Dick (1997). 'Mixing metaphors: a politics or economics of knowledge?' Conference paper. Actor-Network and After. Keele University, U.K., July.

PENCE-AR (2000). Protein Engineering Network of Centres of Excellence Annual Report. Ottawa: NCE Secretariat.

Phillips, Susan D., and Michael Orsini (2002). *Mapping the Links: Citizen Involvement in Policy Processes*. Canadian Policy Research Network. Discussion Paper No. F21 (April).

Phillipson, Donald (1983). 'Steacie myth and the institutions of industrial research.' *Scientia Canadensis* 7 (25).

Phillipson, Donald (1991). 'Building Canadian science.' *Scientia Canadensis* (special issue).

Pickering, Andrew (1993). 'The mangle of practice: agency and emergence in the sociology of science.' *American Journal of Sociology* 9: 559–89. Reprinted 1999 in Mario Biagioli, ed., *The Science Studies Reader*, 372–93. New York: Routledge.

Polanyi, Michael (1940). 'The rights and duties of science.' In Michael Polanyi, *The Contempt of Freedom: The Russian Experiment and After*, 116 pages. London: Watts; reprinted 1975, New York: Arno.

Polanyi, Michael (1962). 'The republic of science: its political and economic theory.' *Minerva* 1 (1): 54–73.

Polster, Claire (1993). 'Compromising positions: the federal government and the reorganization of the social relations of Canadian academic research.' PhD diss., York University, Toronto.

Polster, Claire (1998). 'From public resource to industry's instrument: reshaping the production of knowledge in Canada's universities.' *Canadian Journal of Communication* 23: 91–106.

Porter, John (1965). *The Vertical Mosaic: An Analysis of Social Class and Power in Canada*. Toronto: University of Toronto Press.

Porter, Theodore M. (1995). *Trust in Numbers: The Pursuit of Objectivity in Science and Public Life*. Princeton, NJ: Princeton University Press.

Power, M. (1995). *The Audit Society: Rituals of Verification*. Oxford: Oxford University Press.

Prater, Michael, and Shawn Newlands (1999). 'Molecular genetics and otolaryngology.' Grand rounds presentation; series editor Francis B. Quinn Jr. UTMB Dept of Otolaryngology. University of Texas, Medical Branch, Galveston. Accessed at www2.utmb.edu/otoref/Grnds/Mol-genetics-9910/Mol-gen-9910.htm, 13 October.

Prelli, L.J. (1997). 'The rhetorical construction of scientific ethos.' In Randy Allan Harris, ed., *Landmark Essays on Rhetoric of Science*, 87–104. Mahwah, NJ: Lawrence Erlbaum.

Price, Derek J. de Solla (1963). *Little Science, Big Science*. New York: Columbia University Press.

Pullen, J.W. (1990). *Centres of Excellence*. Report prepared by the Canadian Centre for Management Development. Government catalogue no. SC93–2/2–1990E. Ottawa: Minister of Supply and Services.

Rahm, Dianne (1994). 'University-firm linkages for industrial innovation.' Paper prepared for the Center for Economic Policy Research/American Association for the Advancement of Science conference. University Goals, Institutional Mechanisms, and the 'Industrial Transferability of Research.'

Rappert, Brian, and Andrew Webster (1997). 'Regimes of ordering: the commercialization of intellectual property in industrial-academic collaborations.' *Technology Analysis and Strategic Management* 9 (2): 115–30.

Reid, John R. (2000). *Annual Report, 1999–2000*. Report of the Information Commissioner of Canada. Ottawa: Minister of Public Works and Government Service.

Reid, John R. (2001). *Annual Report, 2000–2001*. Report of the Information Commissioner of Canada. Ottawa: Minister of Public Works and Government Service.

Rip, Arie (1990). 'Implementation and evaluation of science and technology programs and priorities.' In Susan E. Cozzens, Peter Healey, Arie Rip, and John Ziman, eds, *The Research System in Transition* 57: 263–80. NATO ASI Series D: Behavioral and Social Sciences. Dordrecht: Kluwer Academic Publishers, in cooperation with NATO Scientific Affairs Division.

Rip, Arie (1997). 'A cognitive approach to relevance of science.' *Social Science Information* 36 (4): 95–110.

Rip, Arie (2000). 'Fashions, lock-ins, and the heterogeneity of knowledge production.' In Merle Jacob and Tomas Hellstrom, eds, *The Future of Knowledge Production in the Academy*, 28–39. Buckingham: Open University Press.

Rip, Arie (2001). 'Regional innovation systems and the advent of strategic science.' *Journal of Technology Transfer*, special issue on regional innovation systems.

Rose, N. (1996). 'Governing "advanced" liberal democracies.' In A. Barry, T. Osborne, and N. Rose, eds, *Foucault and Political Reason: Liberalism, Neoliberalism, and the Rationalities of Government*, 37–64. Chicago: University of Chicago Press.

Rosenberg, Nathan (1998). 'Uncertainty and technological change.' In Dale

Neef, Anthony Siesfeld, and Jaqueline Cefola, eds, *The Economic Impact of Knowledge*, 17–34. Resources for the knowledge-based economy. Boston, MA: Butterworth Heinemann.

Rubenson, Kjell, and Hans G. Schuetze (2000). *Transition to the Knowledge Society: Policies and Strategies for Individual Participation and Learning*. Vancouver, BC: UBC Press for Human Resources Development Canada (HRDC) and Institute for European Studies, UBC.

Ruggie, John Gerard (1975). 'International responses to technology: concepts and trends.' *International Organization* 29 (3) (Summer): 557–83.

Ruivo, Beatriz (1994). '"Phases" or "paradigms" of science policy?' *Science and Public Policy* 21 (3): 157–64.

Sabatier, P.A. (1988). 'An advocacy coalition framework of policy change and the role of policy-oriented learning therein.' *Policy Sciences* 29: 129–68.

Savoie, Donald J. (1995). 'What is wrong with the new public management?' *Canadian Public Administration* 38 (1) (Spring): 112–21.

Segerstrale, Ullica, ed. (2000). *Beyond the Science Wars: The Missing Discourse about Science and Society.* Albany, NY: SUNY Press.

Shapin, Steven (1994). *A Social History of Truth: Civility and Science in 17th-Century England*. Chicago: University of Chicago Press.

Shapin, Steven (1995a). 'Cordelia's love: credibility and the social studies of science.' *Perspectives on Science* 3: 255–75.

Shapin, Steven (1995b). 'Here and everywhere: sociology of scientific knowledge.' *Annual Review of Sociology* 21: 289–321.

Sheehan, Helena (1993). *Marxism and the Philosophy of Science: A Critical History.* London: Humanities Press International.

Simmel, G. (1950). *The Sociology of Georg Simmel*. Glencoe, IL: Free Press.

Slaughter, Sheila (1998). 'Federal policy and supply-side institutional resource allocation at public research universities.' *Review of Higher Education* 21 (3): 209–44.

Slaughter, Sheila, and Larry L. Leslie (1997). *Academic Capitalism: Politics, Policies, and the Entrepreneurial University.* Baltimore: Johns Hopkins University Press.

Slaughter, Sheila, and Gary Rhoades (1990). 'Renorming the social relations of academic science: technology transfer.' *Educational Policy* 4 (4): 341–61.

Sonnert, Gerhard, and Harvey Brooks (2000). 'The basic-applied dichotomy in science policy: lessons from the past.' In Lewis Branscomb, Gerald Holton, and Gerhard Sonnert, eds, *Science for Society: Cutting-Edge Basic Research in the Service of Public Objectives. A Blueprint for an Intellectually Bold and Socially Beneficial Science Policy,* Appendix A, 50–76. Conference report and consensus document. Conference on Basic Research in the Service of Public Objectives, Washington, DC, November. New York: Columbia University, Center for Science, Policy and Outcomes.

Star, Susan Leigh, and James R. Griesemer (1989). 'Institutional ecology, "translations" and boundary objects: amateurs and professionals in Berkeley's Museum of Vertebrate Zoology, 1907–39.' *Social Studies of Science* 19: 387–420.

Starr, Paul (1988). 'The meaning of privatization.' *Yale Law and Policy Review* 6: 6–41.

Statistics Canada (1999). *Survey of Intellectual Property Commercialization in the Higher Education Sector, 1998 by Michael Bordt and Cathy Read*, 88F0006XPB No. 01; ST-00–01 Cong.

Statistics Canada (2003a). 2001 Survey of Intellectual Property Commercialization in the Higher Education Sector. Preliminary Distribution 1. 6 November 2002.

Statistics Canada (2003b). 2001 Survey of Intellectual Property Commercialization in the Higher Education Sector. Preliminary Distribution 2. 4 April 2003.

Steacie, E.W.R. (1965). *Science in Canada. Selections from the Speeches of E.W.R. Steacie*. Ed. J.D. Babbitt. Toronto: University of Toronto Press.

Stephan, P.E. (1996). 'The economics of science.' *Journal of Economic Literature* 34 (September): 1199–235.

Stokes, Donald E. (1995). '"Science: the endless frontier" as a treatise.' Conference paper. Science, the Endless Frontier 1945–1995. Learning from the Past, Designing for the Future. Columbia University, Part I – 9 December 1994.

Stokes, Donald E. (1997). *Pasteur's Quadrant: Basic Science and Technological Innovation*. Washington, DC: Brookings Institution Press.

Strauss, Anselm L. (1982). 'Social worlds and legitimation processes.' *Studies in Symbolic Interaction* 4: 171–90.

Thistle, Mel W. (1966). *The Inner Ring: The Early History of the National Research Council of Canada*. Toronto: University of Toronto Press.

Toulmin, Stephen (1990). *Cosmopolis: The Hidden Agenda of Modernity.* Chicago: University of Chicago Press.

Toulmin, Stephen (2001). *Return to Reason.* Cambridge, MA: Harvard University Press.

Urry, John (1998). *Locating HE in the global landscape*. Working paper, December. Society for Research in Higher Education. Lancaster University: Department of Sociology.

van der Meulen, B., and A. Rip (1996). 'The postmodern research system.' *Science and Public Policy* 23 (6): 343–52.

Veblen, Thorstein (1918). *The Higher Learning in America: A Memorandum on the Conduct of Universities by Business Men*. New York: Kelley, 1965.

Webster, Andrew, and Kathryn Packer (1995). 'Patents and technology transfer in public sector research: the tension between policy and practice.' Accessed at http://spsg.com/papers/web2.htm#1, in *Science Policy Support Group Working Papers*, 24 September 1997.

Webster, Andrew, and Kathryn Packer (1996). 'When worlds collide: patents in public sector research.' In H. Etzkowitz and L. Leydesdorff, eds, *Universities and the Global Knowledge Economy*, 47–59. London: Cassell.

Webster, Andrew, and Kathryn Packer, eds (1996). *Innovation and the Intellectual Property System*. London: Kluwer Law.

Weiner, Charles (1986). 'Universities, professors, and patents: a continuing controversy.' *Technology Review* 35 (February/March): 33–43.

Weiner, Charles (1989). 'Patenting and academic research: historical case studies.' In Vivian Weil and John W. Snapper, eds, *Owning Scientific and Technical Information: Value and Ethical Issues*, 87–105. New Brunswick, NJ: Rutgers University Press.

Weintraub, Jeff (1997). 'The theory and politics of the public/private distinction.' In Jeff Weintraub and Krishan Kumar, eds, *Public and Private in Thought and Practice: Perspectives on a Grand Dichotomy*, 1–42. Chicago: University of Chicago Press.

Whyte, Anne (1997). '"In the national interest": science, geography, and public policy in Canada.' *Canadian Geographer* 41 (4): 338–49.

Winner, Langdon (1993). 'Upon opening the black box and finding it empty: social constructivism and the philosophy of technology.' *Science, Technology, and Human Values* 18 (3) (Summer): 362–78.

Wolfe, David (2000). 'Globalization, information and communication technologies and local and regional systems of innovation.' In Kjell Rubenson and Hans G. Schuetze, eds, *Transition to the Knowledge Society: Policies and Strategies for Individual Participation and Learning*. Vancouver: UBC Press for Human Resources Development Canada (HRDC) and Institute for European Studies, UBC.

Wulf, William A. (1993). 'The collaboratory opportunity.' *Science* (13 August): 854–5.

Wynne, Brian (1999). 'Expert construction of lay epistemics.' Presidential plenary: The Participation of Lay People in the Production and Dissemination of Knowledge. Society for the Social Studies of Science (4S). San Diego, 27 October – 2 November.

Yearley, S. (2000). 'Making systematic sense of public discontents with expert knowledge: two analytical approaches and a case study.' *Public Understanding of Science* 9: 105–22.

Zucker, Lynne G., and Michael R. Darby (1997). 'Individual action and the demand for institutions.' *American Behavioral Scientist* 40 (4) (February): 502–14.

Index